AF885685

The Philosophy of Time

The question of the existence and the properties of time has been subject to debate for thousands of years. This considered and wide-ranging study offers a contrastive analysis of phenomenologies of time from the perspective of the problematics of the visibility of time. Is time perceptible only through the veil of the change of events? Or is there a naked presence of 'time itself'? Or has time always effaced itself?

McLure's new work also stages confrontations between phenomenology of time and analytical philosophy of time. In doing so he explores from fresh perspectives such ancient issues such as whether time passes, whether experiential time is 'real time', whether the very concept of time is self-contradictory, or subjective.

The Philosophy of Time: Time before times is essential reading for students and teachers of philosophy worldwide.

Roger McLure is a Lecturer in French at Keele University, Staffordshire. He has published widely on topics in philosophy, linguistics and literature.

Routledge studies in twentieth-century philosophy

1 **The Story of Analytic Philosophy**
Plot and heroes
Edited by Anat Biletzki and Anat Matar

2 **Donald Davidson**
Truth, meaning and knowledge
Edited by Urszula M. Zeglén

3 **Philosophy and Ordinary Language**
The bent and genius of our tongue
Oswald Hanfling

4 **The Subject in Question**
Sartre's critique of Husserl in *The Transcendence of the Ego*
Stephen Priest

5 **Aesthetic Order**
A philosophy of order, beauty and art
Ruth Lorland

6 **Naturalism**
A critical analysis
Edited by William Lane Craig and J.P. Moreland

7 **Grammar in Early Twentieth-Century Philosophy**
Richard Gaskin

8 **Rules, Magic and Instrumental Reason**
A critical interpretation of Peter Winch's philosophy of the social sciences
Berel Dov Lerner

9 **Gaston Bachelard**
Critic of science and the imagination
Cristina Chimisso

10 **Hilary Putnam**
Pragmatism and realism
Edited by James Conant and Urszula Zeglén

11 **Karl Jaspers**
Politics and metaphysics
Chris Thornhill

12 **From Kant to Davidson**
The idea of the transcendental in twentieth-century philosophy
Edited by Jeff Malpas

13 **Collingwood and the Metaphysics of Experience**
A reinterpretation
Giuseppina D'Oro

14 **The Logic of Liberal Rights**
A study in the formal analysis of legal discourse
Eric Heinze

15 **Real Metaphysics**
Edited by Hallvard Lillehammer and Gonzalo Rodriguez-Pereyra

16 **Philosophy After Postmodernism**
Civilized values and the scope of knowledge
Paul Crowther

17 **Phenomenology and Imagination in Husserl and Heidegger**
Brian Elliott

18 **Laws in Nature**
Stephen Mumford

19 **Trust and Toleration**
Richard H. Dees

20 **The Metaphysics of Perception**
Wilfrid Sellars, critical realism and the nature of experience
Paul Coates

21 **Wittgenstein, Austrian Economics, and the Logic of Action**
Praxeological investigations
Roderick T. Long

22 **Ineffability and Philosophy**
André Kukla

23 **Kant, Cognitive Metaphor and Continental Philosophy**
Clive Cazeaux

24 **Wittgenstein and Levinas**
Ethical and religious thought
Bob Plant

25 **The Philosophy of Time**
Time before times
Roger McLure

The Philosophy of Time

Time before times

Roger McLure

Routledge
Taylor & Francis Group
LONDON AND NEW YORK

First published 2005
by Routledge
2 Park Square, Milton Park, Abingdon, Oxon OX14 4RN

Simultaneously published in the USA and Canada
by Routledge
270 Madison Ave, New York, NY 10016

Routledge is an imprint of the Taylor & Francis Group

© 2005 Roger McLure

Typeset in Garamond by Wearset Ltd, Boldon, Tyne and Wear
Printed and bound in Great Britain by MPG Books Ltd, Bodmin

All rights reserved. No part of this book may be reprinted or reproduced or utilized in any form or by any electronic, mechanical, or other means, now known or hereafter invented, including photocopying and recording, or in any information storage or retrieval system, without permission in writing from the publishers.

British Library Cataloguing in Publication Data
A catalogue record for this book is available from the British Library

Library of Congress Cataloging in Publication Data
A catalog record for this book has been requested

ISBN 0–415–33178–1

To the memory of my father, Capt. M.J. McLure, for the grain of salt that sets the cussed course; to Prof. J.H. Broome, for the spacious days in which to think; to Whyeda for her unfathomable responses.

Contents

Note on the text

Every effort has been made to obtain permission to reprint material in this book. The publisher would be grateful to hear from any copyright holder who is not here acknowledged and will undertake to rectify any errors or omissions in future editions of this book.

Introduction

In Part I of this book I have tried to contrast three phenomenologies of time and, in Part II, to stage polemical confrontations between phenomenology of time generally and analytical philosophy of time. I will let the chapters themselves justify my choice of phenomenologists, namely Bergson, Husserl and Levinas. All that might be remarked here is that the pattern of contrasts within phenomenology of time is rich enough to make any choice of exponents potentially as good as any other. The omissions enforced by choice (by limitations of space) are, however, partly compensated by the presence in Part I of phenomenological voices critical of my chosen trio, and by Part II offering an open forum to all phenomenologists with something of interest to say about the ecumenical time-problems considered therein.

One filter selecting my material on the exegetical side has been the desire to pitch contrasts at the level of the time-*models* on offer in phenomenology, a focus which closes off considerations that would be structural to a book tracing personal philosophical itineraries, influences, developments, changes of mind, etc. I have concentrated on the mature positions of the philosophers concerned (though, in the case of Bergson, some dalliance in the direction of philosophical biography served the purposes of exposition). In other words, the historical phenomena are minimally historicized. This method is akin to the linguistician's practice of describing different historical states of a language-system synchronically, that is, at instants cut out from the processes of change in which the states are really involved. Without these 'snapshots' the stable data needed for contrasts would remain unavailable. It goes without saying that this approach is not in competition with the historicizing approach characteristic of most Continental philosophical scholarship.

The background of consensus against which contrasts stand out is strong enough to suggest that the development of the phenomenology of time has been driven by family quarrels and skeletons beneath alleged 'foundations': the quarrel picked with Bergson's naturalism by Merleau-Ponty and Sartre (Chapter 1); Heidegger's and Sartre's subordination of time to very different concepts of ontology; the rejection of the fundamentalness of previous phenomenological times entailed by Levinas's pre-intentional (yet non-Bergsonian), de-ontologized (yet non-Husserlian) conception of time as fundamentally 'ethical' (Chapter 3). These differences justify the task of exegesis because they are much more elusive than the similarities deriving from consensus over such basic principles as the unrevisability of the immediate 'data' of experience, the experiential as distinct from linguistic sources of meaning, the independence of time from events.

The main bone of contention among phenomenologists has not been who is right or wrong in an absolute sense, but rather who has taken the prize of *fundamental* ("original") time. Phenomenologists have in consequence been willing to grant validity at a non-fundamental level to cousinly time-theories whose claim to fundamentalness they reject. Thus, although Levinas and Heidegger deny, against Bergson and Husserl, that fundamental time passes, they do not deny that passage characterizes derivative times. In this as in other respects, phenomenologists have tended to *distance* themselves from each other, in the game of 'being more fundamental than thou', rather than rejecting the other's view outright.

A fundamental time is foundationalist only when construed as framing presence. Where it does not frame presence, the concept of the absolute past moves centre-stage, frustrating the original phenomenological aspiration to render time visible, present. (The concept of the absolute past seems to be wholly absent from analytical philosophy of time.)

My main exegetical thesis is that all phenomenologies of time have in common a conception of time as independent of non-temporal factors such as spatial events (changes), the time-points associated with these, even psychological content. This common claim I dub 'the autonomy thesis'. Though it goes back at least as far as Augustine (see, p. 176), it was re-discovered by Bergson, from whom phenomenology inherited it (ungratefully, as a family heirloom). This legacy of Bergson's is my justification for treating him as a phenomenologist, though a much wider range of reference would be necessary to settle the question of how far his philosophy as a whole anticipates phenomenology.

Judging by the collected phenomenological testimonies, the autonomy thesis does not entail the thesis of time-motion (the passage of time). But the thesis of time-motion does presuppose the autonomy thesis, in a way I shall now try to suggest with specific reference to the time of consciousness.

It will be granted that we make sense of such statements as 'My future is what my present will become', 'My past is what my present used to be.' In this sense of 'present', as the present of consciousness, a current present acknowledges a past present as what it used to be, or equivalently, a past present acknowledges a current present as what it has become. These mutual acknowledgements, made across changes of mode-of-being (from e.g. *is* to *used to be*) are the self-synthesizings through which consciousness 'constitutes' its temporal unity. We have here a discernible dimension of 'pure time' and an intelligible sense in which it is passage (into itself), the production of its own unity. Without that unity Hume could not even have collected the "bundle of sensations" he pronounced unworthy of the name 'self'. Now, clearly, none of this holds once presentness has been logically linked to events. Though it might please us to suppose that past events somehow survive, the mind baulks at the thought that they could survive as what present events used to be, or that present events are what past events have become. If *time* passes it must do so independently of the passage of events; for time does not pass away; and events, which do pass away, do not take on a new lease of life as what present events used to be.

Since the 'time itself' of phenomenology and the time of Newton have shared the label 'absolute time', it is perhaps worth making clear how they differ. Newton's time is a mathematical continuum of time-slots for events which logically need not fill them. There would still be time even if there were no events, because there would still be times: time requires only times. Phenomenological time, on the other hand, is prior to times (as my subtitle announces), as well as to the events occupying (or, more likely, generating) them.

The events regarded as inessential to time are ordinary natural events: the bark of a dog, the positional events comprising a physical motion. As against these common-or-garden events, Heidegger and Levinas invoke an esoteric sense of event, which they link essentially with time: for Heidegger, time is somehow an 'event of Being', for Levinas, it is the ethical 'event beyond Being' that opens the self onto the Other. In both cases we are dealing with time as event rather than as ordered structure. The point worth stressing, then, is that the autonomy thesis is compatible with wedding time to 'events' of a *non-natural* sort.

What 'is it' that is left over after the subtraction of natural events? What is the 'is' of time? Certainly, no phenomenologist takes the autonomy thesis to imply that time is some sort of *entity*-in-itself: if it is a substance in the sense of 'needing nothing further in order to exist', it is not a substance in the sense of a natural thing or process. As I try to show in Chapter 4, pp. 136–138, phenomenologists (at least post-Bergson) would agree with Wittgenstein that time is not a "queer process", while rejecting his corollary that there is 'nothing there' to be baffled about (just linguistic muddles to clear up). There *is* something 'there' and it is queer; but it is queer because it is not a process and not because it is a queer process.

This makes the 'is' of phenomenological time even queerer than the 'is' of ordinary 'eventful' time, which is semi-naturalized by the empirical 'isness' of the events it relates (as Heidegger points out, see p. 192). As I indicated, phenomenology of time has aspired both to twin itself with fundamental (pre-categorical) ontology and to go 'beyond' all ontology. In both cases time assumes a status that cannot be captured by the ordinary empirical 'is' of categorical ontologies; yet this 'is' continues to be used, playfully or abusively depending on one's point of view. Placing fundamental time beyond Being might seem like an excellent opportunity to ditch 'is' altogether. But of course Levinas doesn't, he can't.

The idea of staging the problem-centred confrontations of Part II grew out of dissatisfaction with the symmetrical limitations of both traditions:

analytical philosophy's neglect of pre-linguistic meaning; and phenomenology's sub-propositional stuntedness. There is a need all round to come out of blinkering 'perspectives' to see how any favoured perspective stands in relation to the way time-problems are set up in other traditions. The empirical basis on which they are set up in all strains of analytical philosophy of time, save the scientific, is a stock of troublesome truisms; and that is intrinsic to its value. For even if common sense is the metaphysics of the caveman, it is our natural state of corruption and not to be cured by extended holidaying in the Heideggerian light. The empirical basis of much phenomenology of time is rich and exotic; and that is intrinsic to its different value, which is to let us see this from outside the cave in which we must live.

The topics of Part II sound out phenomenology of time for resources by which it might respond to questions and objections that analytical philosophers would put to it, supposing they bothered with it: the objection, for example, that no phenomenological time could be real time (Chapter 6). In making these responses I have been conscious of pushing phenomenology of time in a direction in which it was not developed by its exponents, squeezing the most out of small bits of texts, even out of passing remarks. This is germane to any attempt to seek out the borderlands between the two traditions.

The oft-remarked methodological differences between analytical philosophy and phenomenology are fully exemplified in their approaches to time: the method of more or less formalized analysis of statements in which temporal words occur, as against the method of more or less transcendental reflection on the experiences in which temporal concepts are supposed to be rooted. Accordingly, analytical philosophy of time interprets the relation between dynamic and static aspects of time in terms of the distinction between tensed and tenseless statements, thereby bringing the metaphysics of time within the semantics of propositional truth. By contrast, phenomenology investigates time at pre-propositional levels, appealing to sub-linguistic meanings encountered in experience or 'constituted' by consciousness.

The (originally Augustinian) notion that time is something self-dependent is abhorrent to almost all strains of analytical philosophy of time. The contrary analytical (originally Aristotelian) doctrine that time is essentially connected with the fact or perception of change – just a relation between change-events – comes together with the analytical method in the ubiquitous presence of sentences like 'Event *e* is present/past/future' and 'Event *e* is tenselessly earlier than event *f*'. No equivalent sentences are to be found in phenomenological philosophy of time.

The notion that time passes is no less of a heresy in the analytical camp (though the sport it has afforded some philosophers has been at the expense of everyday English locutions rather than of phenomenology). As understood by phenomenologists, the proposition that time passes is not the innocuous one that 'events come and go', or more technically, that tensed sentence-tokens have irreducibly tensed truth conditions: it is the proposition that 'time itself', as something prior to events, 'passes'. I defend this thesis of time-motion in Chapter 4.

Though all phenomenologists implicitly deny the basicness of the tensed vs tenseless distinction, the unique relevance of Husserl to McTaggart's paradox (Chapter 5) derives from his canvassing a concept of *serial* time broadly commensurable, in a way the non-serial times of Bergson and Levinas are not, with the exclusively serial times considered by McTaggart and

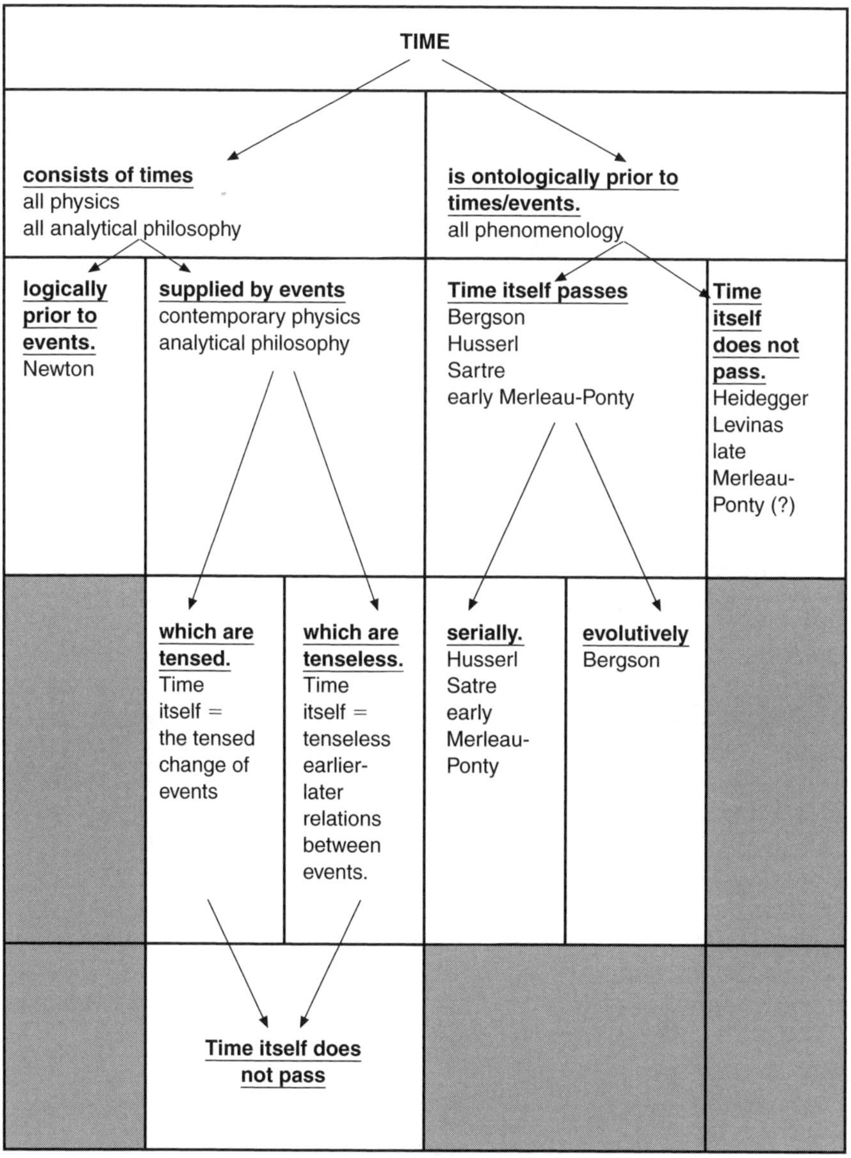

Figure 1 The constellation of positions sketched by the exegetical claims of this book.

mainstream analytical philosophy. Since Husserl, the two traditions have tended in divergent directions, the one upwards from experience towards greater logico-linguistic sophistication, the other downwards to expose ever more exotic currents of time-experience, territory from which the traditional time-paradoxes are simply invisible. But even Husserl's ideas on time make contact with analytical philosophy only after considerable upward-pushing.

It is the conclusion of this book, stated here once and for all, that the two traditions of philosophy of time have moved ever further beyond the range of rapprochement, and that notwithstanding the chummier atmosphere now prevailing, the gap between them is not going to yield to pressure from transatlantic brokerage (*pace* Paul Ricoeur). In saying a moment ago that their methods differ, I allowed myself an understatement which neglects the fact that the difference of method reflects the different elements in which the two traditions work: the appropriate analogy is the difference, not between working the soil with a spade and with a fork, but between working the soil and working the sea.

For one simple reason this book has taken an unmentionable length of time to complete – the sheer difficulty of deciding exactly what these phenomenologists are saying: months spent following false exegetical trails, pondering sentences armoured in shining obscurity. Even so, these texts teem with insights, beside which the 'temporal facts' of analytical philosophy appear jejune. I imagine my ideal readers to be people who, having struggled with this philosophy and seen through a glass darkly, want to see more clearly.

All the chapters may be read as self-contained essays – most of them began as such. The translations from the French and German are my own.

Part I

Contrasts

1 Time as creative process
Bergson

Introduction

Some three decades ago Frederick Copleston summarized the estimate of Bergson[1] in the English-speaking world and continental Europe as follows:

> Although Bergson once had a great name, his use of imagery and metaphor, his sometimes rather highblown or rhapsodic style, and a certain lack of precision in his thought have contributed to his being depreciated as a philosopher by those who equate philosophy with logical or conceptual analysis and who attach great value to precision of thought and language . . . In some countries, including his own, he has fallen into neglect for another reason, namely the eclipse of the philosophy of life by existentialism and phenomenology.[2]

No useful purpose would be served by dwelling on the charge of lack of precision (justified though it is).[3] More instructive, because germane to Bergson's intuitive method, are his own ideas about "precision in philosophy" (PM, 1–3), in particular his distinction between approaching philosophical problems through ready-made multi-purpose concepts or logical laws and approaching them through intuitions which are precise inasmuch as "tailored to the reality in which we live" (PM, 30). Intuition here means insight into the natures of things, eliciting *their* particular 'logic'. On the intuitive standard of precision, a logically precise expression of a state of affairs that is not logically structured would be fundamentally imprecise – applying the law of the excluded middle to time, for example. The trouble with the metaphysical systems of the past, Bergson tells us, was that they violated the first principle of precision, which is that "the explanation that we must deem satisfactory is one that adheres to its object" (PM, 1). Bergson is not against metaphysics: his target is rather the arbitrary synthesizing that has constructed metaphysical systems baggy enough to accommodate possible worlds very different from our real world.

One might of course deny the existence of anything like Bergsonian intuition. But then the issue between Bergson and his rationalist (analytical) detractors switches to what should count as "true empiricism" (PM, 196).

While Bergson's statement that "a conception is worth no more than the possible perceptions it represents" (PM, 145) might sound like a verificationist slogan in a non-linguistic key, any rationalist verificationist would withdraw sympathy on discovering that by 'perceptions' Bergson means 'intuitions'; whereby it becomes clear that this verificationism owes allegiance less to empiricism than to *le sensualisme anglais* – French for the metaphysical doctrine that the empirical is exhausted by the sensory. According to the letter of standard (British) empiricism, all seeing is sight, insight being restricted to logical matters. Bergson has a less stingy concept of empiricism, allowing for empirical insight (intuition).

As for the role of existentialism in eclipsing Bergsonism in France: the existentialist revival of the Hegelian motifs of negativity took the gloss off Bergson's famous arguments against 'nothingness', while the standard interpretation of his concept of intuition as a fusion of self with world seemed to many people to inflate the self to a fullness beyond due measure (Sartre slighted Bergsonism as a "philosophy of fullness"). Fullness, moreover, did not seem to sit comfortably with the themes of 'trace', original surplus and belatedness, subsequently ushered in by Derrida's critique of 'the metaphysics of presence'.[4] Justified though it is by the surface of Bergson's texts, the view of his intuition as fusion has been challenged by Gilles Deleuze's reconstruction of Bergsonism around a new conception of temporal difference. We shall catch this at work at several points later on. Here it will suffice to indicate that the Deleuzean approach to Bergson sees him as making a seminal contribution towards the philosophy of difference, by dissociating difference from negation: "the essential feature of Bergson's project is to think difference independently of every form of negation".[5]

(These purely philosophical considerations would have to be supplemented, in any cultural history of French philosophy, by mention of the perceived irrelevance of Bergson's luxurious metaphysical vision and perhaps facile optimism to the deflationary moral atmosphere of war-weary Europe. The sensitivity of intellectual life in France to the broader socio-historical context, its proneness to fashion, are never negligible factors in the fates of French philosophers.)

As for phenomenology's part in backstaging Bergson: although French phenomenologists have tended to be existentialists, classical Husserlian phenomenology has a technical orthodoxy, deriving partly from Kant, which is independent of the (quasi-Hegelian) doctrines to which some of the existentialists harnessed it. And it is this technical orthodoxy, rather than the existentialist doctrines, which is invoked by Merleau-Ponty and Sartre in their influential criticisms of Bergson. Where Bergson had thought against Kant, French phenomenology inherited via Husserl, along with the latter's intentional theory of consciousness, some of the basic tenets of the Kant's non-naturalistic concept of mind (synthesis, transcendentalism, constitution, etc.). Against these the naturalism Bergson was measured and found wanting. To exhibit this Kantian-phenomenological critique is my main contrastive purpose in this chapter.

A more comprehensive view of the French phenomenologists' estimate of Bergson would show some of them to be appreciative of him as a precursor of phenomenology, anticipating some phenomenological doctrines and creating, through his reaction against positivism, the climate in which phenomenology could flourish in France. While Sartre never softened his orthodox strictures, other French phenomenologists, less orthodox or in their less orthodox periods, have stressed Bergson's anticipations of Heidegger. Thus Levinas:

> Bergson's theory of time as concrete duration is . . . one of the most significant, if largely ignored, contributions to contemporary philosophy. Indeed, it was this Bergsonian emphasis on temporality that prepared the soil for the subsequent implantation of Heideggerian phenomenology in France.[6]

And the later (as distinct from earlier and critical) Merleau-Ponty is more specific on the same point:

> the original perception which we re-discover in ourselves and that which shows through in evolution as its inner principle are laced into each other . . . , we are always dealing with the same tension between one *durée* and another *durée* which frames it from the outside.[7]

Of these tensions or temporal rhythms more anon. The point here is just that Bergson is being credited with something like a home-produced prototype of Heidegger's idea of time as a horizon for the understanding of Being: "it is now being as a whole that has to be approached from the point of view of time",[8] says Merleau-Ponty of the Bergsonian framing of human time by a temporally-defined external reality. A further consideration here is that Bergson's battle-cry to philosophy to rediscover its 'stone' in "the immediate data of consciousness" had been heeded by French philosophers several decades before they heard it echoed in Husserl's slogan "back to things themselves".[9]

Since Copleston's comment there has been a resurgence of interest in Bergson on the Continent and North America, driven partly by the gradual working-through of Deleuze's influence and partly by the recognition that Bergson's texts offer much that is of relevance to contemporary debates in philosophy of mind, aesthetics and some of the sciences of complexity.[10] My contrastive purpose in this book, however, imposes a narrowing of the focus: to Bergson as just a philosopher of time and again to his philosophy of time to just a phenomenology of individual time-experience, disregarding his theory of cosmological time. The fact that Bergson took his phenomenology of time – his concept of 'duration' or *durée* – as a model for his later theory of cosmological time attests that the former can stand, as it will here, independently of the latter.[11]

Time and substance

From the point of view of the main exegetical thesis of this book, it is the autonomy thesis or concept of 'pure time-change' that constitutes Bergson's most important legacy to phenomenology. For as I intimated in the Introduction, post-Bergsonian phenomenology as a whole appropriates this doctrine under one interpretation or another, shedding most of the accessory doctrines specific to Bergsonism. The autonomy thesis does not affirm that there is time without change but rather that the change essential to time is internal to time itself, not parasitic on non-temporal factors such as objects or change-events in space; there is such an affair as '(the change of) time itself' or 'pure time'. Bergson's version of this is the independence of time from space.

However, if the ordinary things (substances) of our world are bound to space, and if space is external to time as Bergson claims, then it might seem that Bergson's time-change (*durée*) is something wholly immaterial, and opposed to a domain of space-bound timeless substances. That this is his view appears to be confirmed by an unusually explicit text: the continuity of our inner life, he says, is that

> [of] a flowing or of a passage, but of a flowing and of a passage which are sufficient in themselves, the flowing not implying a thing which flows and the passage not presupposing any states through which one passes: the *thing* and the *state* are simply snapshots artificially taken of the transition.
> (DS, 41)

This change involving no 'thing' would be contentless, the opposite of qualitative change. A self, or anything else, that changed this way could not be a substance. *Durée*, the change of time itself, would involve nothing resembling a process of change in a natural thing, no descriptive language would apply to it. (We shall see that this is precisely how *durée* is interpreted by Husserl and his followers.)

Yet all this conflicts with the theme of the concreteness of time in Bergson. His time is qualitative (a "qualitative multiplicity") as distinct from homogeneous, and "concrete" (DI, 75, 77) as distinct from an abstract form indifferent to the contents passing through it. Bergson belongs to the small band of philosophers who refuse to separate the content of time from its form: "the content is of a piece with its *durée*" (PM, 11). The content? Can *pure* time have a content? Have we caught Bergson in contradiction with himself?

To see why there is no contradiction we need to focus on the word 'thing', which has so far had a free ride. What I think Bergson means by it is conveyed more exactly by the word "support" in his statement that in a heard melody "everything is becoming, but . . . the becoming, being substantial, has no need of a support" (PM, 140–141). Or again, referring to the familiar paradoxes of change: "if our intelligence persists in judging [change] incoherent, adding on I know not what support, it is because . . ." (PM, 8). This support is clearly the occult extra-temporal (thus space-like, unchanging) substratum postulated by Locke and his followers to explain the possibility of a thing coming out of a change as the same thing (in a new state) that went into it. Now I think Bergson's idea is that the change of time *presupposes no non-temporal substratum*, not that it *involves no thing that changes*. It's not that his flow is limpid, so to speak, but that it involves no *non-flowing* thing to which it is relative. The change of time encompasses the change of things, which therefore do not change in space.

It is doubtful whether even common sense believes that the change of inner life consists in some timeless substratum-self donning and doffing properties. That is why Bergson finds in the change-structure of inner life a privileged example of *durée*. An even better example (his favourite) is the experience of listening to a melody, where there is clearly no thing that runs self-identically through the music undergoing change *en route* (cf. a runner who remains the same man throughout his changes of position).[12] Notice that this musical paradigm does not support the idea of an experience of time-structure without content, of a limpid time-flow. The absence of an invariant thing running through the melody is experienced, rather, as the condition of temporal structure *generating* content (sound).

Writing large the musical example, we can say that for Bergson, as against Aristotle and common sense, change (as pure time-change) precedes and determines substance. There are not first things which then accidentally undergo changes; but rather a style of temporal change is essential to a thing and fixes its specific nature. Change is in that sense "substantial": "this change is indivisible, it is substantial" (PM, 8); "the substantiality of the self is its very *durée*" (PM, 76); "its *durée* [life's] is substantial inasmuch as pure *durée*" (PM, 80). The change of a leaf, say, from green to yellow, permits of no distinction between the leafhood of the leaf and its changing, between substance and change: a leaf is a particular complexity of change. So substance, duly distinguished from substratum, has been saved within a philosophy of process (always an achievement): "I am thereby in no way setting aside *substance*. On the contrary, I affirm the persistence of existences. And I believe that I have facilitated their representation" (PM, 211).

An attractive consequence of Bergson's rejection of the substratum theory of substance is that it allows for a holistic conception of change which matches our actual perception of change. Consider: nobody denies that an apple that is red and sweet, having been green and bitter, is an apple that has changed. But ought we to say that the apple has changed *from* green to sweet, and *from* bitter to red? Or ought we not to say that its change to red can be from green only, and its change to sweet from bitter only? My informants are unanimous that the case for second sort of change is somehow stronger, but that the first nevertheless deserves to be accommodated within our concept of change. I think it is clear that the reason why the second sort of change is felt to be more compelling is that, without it, a substratum would be inconsistently qualified: to be both red and green all over an apple *must* change, whereas no change is necessary for something to be bitter and red or green and sweet (witness red peppers and asparagus). There is a temptation, to which Kant but not my informants succumbed, to ignore the first sort of change and treat all change as deducible from the need to avoid contradictory predications (the so-called 'consistency theory of change'). But the temptation holds sway only as long as we are induced (by *inter alia* the language of discrete properties and from–to pairings) to picture change in terms of the substratum theory, for only then is it possible to distinguish between the two sorts of change and favour the one over the other.

The well-known difficulties associated with substratum – its existence not being independently identifiable (PM, 211), how the properties it 'supports' really relate to it – would be good reasons for ditching the notion. But can we do without it? Well, the substratum is supposed to have two functions: to *unify* properties and from–to pairings which, being essentially discrete, would otherwise fall apart; and to explain the *identity* of things across change. The unifying function is rendered redundant, however, if we construe things as synthetic wholes where every quality is conditioned by every other, as in this lemon of Sartre's: "the lemon is extended throughout its qualities and each of its qualities is extended throughout every other. It is the bitterness of the lemon that is yellow, it is the yellowness of the lemon that is bitter."[13] Such synthetic wholes contrast with congeries of discrete properties, any one of which may change while the others do not: the former change as wholes *into* other wholes, producing a "continuity of sensory qualities" (MM, 222). In virtue of their synthetic connectedness, the qualities of the lemon hold *themselves* together, dispensing with the need for a unifying substratum. We might still seem to be left with the metaphysical problem of the preservation of identity through change: what makes a holistically changed lemon the same lemon as before the change? Bergson's answer is that the problem is false, in that it wrongly assumes that a thing is what it is independently of change, landing us with the problem of explaining how it remains what it is *despite* changing: "the permanence of substance is in our view a continuity of change" (PM, 96).

The self-world relation as temporal difference

In the Foreword to his second work, (MM, 1896), Bergson announces that "*questions concerning the subject and the object, their distinction and union, should be asked in terms of time and not in terms of space*" (MM, 74; italics original). If this is true, the subject–object relation holds within time. Space lies outside that relation, outside time. The object, no less than the subject, is temporal without being *spatio*-temporal. Time embraces objects (matter) while remaining independent of space. To elucidate this version of the autonomy thesis is my main task in this section.

In Bergson's first work (DI, 1889) the treatment of time is undertaken for the sake of a defence of freedom. Time is crucial to his main argument, but this is not an argument about time. The argument for freedom applies Bergson's general methodological principle of unmixing ontologically non-immediate (therefore philosophically treacherous) 'mixtures'; in the present case the mixture of time and space on which, he claims, rest both the fallacies of the determinists and the false conception of freedom advocated by their libertarian opponents. (Note that this mixture always consists of spatialized time rather than of temporalized space: we speak of 'lengths of time', etc.) What matters for the defence of freedom is less that time be fully and unambiguously characterized than that it be shown, by this strategy of unmixing, to be under any characterization opposed to space. With this subordination of time to freedom in DI Bergson initiated his lifelong practice of developing his concept of time as a by-product of the specific philosophical tasks he sets himself from book to book, only the last of which (DS) has time as its main theme.

In a formulaic statement DI affirms the space vs time opposition obtained by the method of unmixing: "outside us there is mutual exteriority without succession; inside us there is succession without mutual exteriority" (DI, 95). But this is saying that the distinction between the subject and the object is precisely not a distinction within time, but between time and space. While the difference of essence between time and space will remain a constant of Bergson's thought, his views on how the distinction is specified in relation to the 'inside' and the 'outside' will change radically. The subject–object distinction will be rethought as holding between degrees of temporal difference (yet that difference will not be merely a difference of degree!).

Why does Bergson demur from temporalizing matter in DI? The answer the book gives us seems to be that doing so would upset the deterministic principles of mechanics, since these rest on the assumption of the conservation of a fixed quantum energy at odds with the creative character Bergson attributes to *durée*. His extension of *durée* to the material world, from MM onwards, implies a less deferential attitude towards the conservation principle (in tune with quantum mechanics) rather than any compromise on *durée*'s creativity. It also registers his attempt to attenuate the stark dualism implicit in the formulaic statement: how could the inside and the outside meet, once they have been defined by mutual exclusion?

In DI, time is identified with consciousness, in MM, with mind and with 'memory' in the "special sense" (MM, 148) Bergson gives to the word. We need to get a working understanding of this memory before we can approach the role of time in the solution of the self-world problem.

One species of memory is the motor-memory evidenced in unconscious habitual behaviour and involving no imaging of the past. It is in this sense of memory that my reciting a poem by heart is evidence of my having remembered it. According to Bergson, this motor-memory has automatized the past and is metaphysically part of matter. But we remember in a second sense which is apparently independent of the first, as the first is of it: with or without imaging, I can remember having learnt a poem (where, when and how), whether or not I also have the motor-memory of it. Neither of these memories is memory in the special *ontological* sense Bergson has in mind. This memory is internal to time itself, it is not an episode in the life of the body, or a faculty of mind we occasionally exercise. Transcendentally, though not phenomenologically, this memory is prior time: there is time because this memory articulates it.

Being identified with mind, this memory is one side of the self–world relation. This means that if Bergson wanted to characterize that relation as involving (more or less straightforwardly) a dyad of substances differing in kind, then mind-memory would have to be defined by counterdistinction with world: each would have just the properties the other lacks. In fact, Bergson does want to uphold a (sophisticated) dualism of substance; which is why he says that mind-memory is "pure", meaning that it is not the memory that is mixed with perceptual elements in concrete experience. Concrete experience is traditionally taken to consist of a blend of perceptions with memories (or general ideas formed from them) which enable us to recognize what is perceptually encountered as this or that sort of thing. These memories are, for Bergson, actualizations of elements of the unconscious mass of pure memory, which is "virtual" (MM, 148) and does not combine with the perceptual elements in experience. Pure memory, then, is not sensorily experienced. Behind the actualized memory (which may be a memory-image) the door leading down to the pure memory whence it emerged has closed. If it lets through no chink of light, the memory-image will bear no trace of its origin and therefore none of the difference in kind of pure memory (mind) from perceptions (matter); for *as actualized*, memories differ from perceptions only in degree (as Bergson acknowledges: MM, 148). So what justification could there be for postulating this pure memory? Bergson claims that pure memory is screened out by our biological orientation towards action; for this turns the attention of consciousness away from the past (from pure mind-memory) and towards the future. Rather than following the progress of the memory out of the past into present consciousness, we "take [the remembered image] as ready-made, as realized in the state of a weak perception [so that] we will close our eyes on the pure memory which that image has progressively developed" (MM, 149). There is a 'light' to be seen but we are biologically programmed to look away from it, towards the future. Were we to follow its progress we would see that the actualized memory

> remains attached to the past by its deep roots and [that], if, once realized, it did not feel the after-effects of its original virtuality, if it were

> not, at the same time as it is a present state, also something contrasting with the present, then we would never recognize it as a memory.
>
> (MM, 148)

But we normally don't follow the progress, we "sacrifice . . . the beginning to the end" (MM, 149), with the result that the actualized memory typically appears as a weak perception. Whereas if we did follow it, we would experience the actualized memory(-image) as an emissary from the past, as "a manifestation of mind" (MM, 271). Bergson seems to be saying that we benefit from the light even as we look away from it; for if we didn't benefit from it, we would constantly be mistaking memories for perceptions; on the other hand, if we did not look away from it, we would not suppose that the memories and perceptions conjoined in concrete experience differ only in degree of force (as Hume supposed). What is emerging here is that the best justification for pure memory is to be gained from the difficulties encountered by the empiricist theory of memory (which is never far beneath the surface of Bergson's discussions of memory). The fact that rememberings are events going on *now*, just as perceivings are, raises the question, What is it about one sensation that tells you it is a memory, about another that tells you it is a present perception? The standard answer offered by the traditional empiricist theory of memory was couched in terms of a distinction of *degree*: images of past were said to be characteristically 'weaker' than perceptions, which "enter with most force and violence" (as Hume said: see Bergson on "strong states and weak states", MM, 149). The unreliability of these quantitative indices – the fact that they would lead to misidentifications far in excess of the actual case – has been well documented[14] and need not detain us here. What matters is just that on this theory our knowledge of the past depends on inferences from present weak sensations, but that the discomfiture of this theory argues that we have a *direct* access to the past, an understanding of the past against the background of which we are able to interpret a subset of sensations as memories. Now, Bergson's pure memory is justified as a possible conception of the possibility of that critical understanding:

> the pure and simple image does not represent to me something past unless it is in fact to the past that I have gone back to fetch it, following the continuous progress which has brought it out of obscurity into the light. . . . The truth is that we shall never reach the past unless we place ourselves directly in it.
>
> (MM, 150, 149–150)

That truth was seen by Bergson long before it was conceded by the empiricist critics of the empiricist theory of memory.[15] What this tells us about the mixture of past and present in concrete experience is that it exists *to be seen through*, by an exercise of intuition which cuts beneath the sensory (psychologically immediate) surface of 'mixed' experience to grasp its concrete (ontologically immediate) conditions. As Deleuze puts it: "Intuition takes us beyond the state of experience towards the conditions of experience. But these conditions are neither general nor abstract, they are not broader than what is conditions, they are the conditions of real experience."[16] If this is transcendentalism, it is of a 'tailored' variety: the conditions are intuitively grasped *within* experience as its pre-sensual underside, thus as conditions of real (as distinct from Kant's merely possible) experience.

Similar remarks apply to the "pure perception" Bergson posits as the polar opposite of pure memory. Pure perception is the limitative concept of perception without memory, just as pure memory is the limitative concept of memory without perception. And just as in pure memory we are immediately placed in the past, so pure perception "places us immediately amidst things" (MM, 70): our perceptions take place *in* things, just as they appear to, not in our heads: they are "originally in things, rather than in the mind, outside of us rather than inside us" (MM, 246) – a proposition which disposes of the 'problem of the external world' as deftly (if at all) as pure memory disposes of the problem of the past. Pure perception (reminiscent of Leibniz's 'momentary mind'), is an impersonal "partial coincidence" (MM, 246, 250) of mind with the system of matter; my *perceptum* stands to the whole of matter as less to more, not as phenomenon to noumenon (MM, 32). Just as pure memory has always receded behind the memory-images as which it is actualized, so pure perception has always-already been interpreted by memory ("*practically we perceive only the past*" (MM, 166–167; emphasis original) and exists "in right rather than in fact" (MM, 31, 68, 246): by which Bergson means, not that pure perception is not really a perception at all, but that it is never really present (sensorily experienced).

There is a deep ambiguity about this pure perception: is it my blind perception of matter, or matter's self-perception? Bergson embraces the ambiguity to claim that the mind is caught in the shadow of matter and vice versa: "the [traditional] opposition between perception and matter is the artificial work of our understanding . . . it is not given to immediate intuition" (MM, 275); "our perception being part of things, things partake in the nature of our perception" (MM, 202). But does this not mean that time-mind participates in the nature of space?

Before answering, let us be clear that what is at stake here is the possibility of extending the autonomy thesis from the time of the self (or consciousness) to the temporal relation between self and world. Is this project not blighted by the participation of perception in matter, considering that matter is spatial and that space is the *other* of time? Yes, if the space of matter is the homogeneous geometrical space that in DI is defined by counterdistinction to time: "space must be defined as homogeneous" (DI, 73). No, if the space of matter can be conceived as timelike and dynamic, as the statement that things (space) partake in the nature of our perception hints it can be.

The latter timelike perception of space is attested throughout MM by a series of U-turns on subsidiary theses of DI. In DI, the intellect was blamed, not for separating perception from matter, but for confusing them; specifically, for importing into sensations the mutual exteriority of the spatial objects which putatively cause them. Or again, and in apparent contradiction to the non-spatial character of sensations affirmed in DI: "*All* sensations [not just tactile ones] participate in extension; all are more of less deeply rooted in extension" (MM, 243) – otherwise it would remain a mystery how originally unextended sensations come to be tied to spatial locations (MM, 242). Extension, moreover, is now said to be indivisible: "nor is what is real an extension divided into independent parts" (MM, 275). And while in DI "[reciprocal] exteriority is the distinguishing characteristic of things which occupy space" (DI, 73), in MM we are told that "the essential characteristic of space is continuity" (MM, 220) – the earlier-mentioned "continuity of [interdependent] sensory qualities". (Let us not jump to the conclusion, however, that Bergson is re-mixing what he began by unmixing, repudiating his dualisms. . . .)

What has happened is that a concept only touched upon in DI has moved centre-stage: "concrete extension", "extensity" (*l'extensif*). This concrete extension, which is lived spatiality, is indivisible, qualitative ("the diversity of sensory qualities" (MM, 244) and mobile – everything geometrical space is not. Given this contrast, Bergson can say that sensations are *extended* without recanting the doctrine of DI that they are not geometrically *spatial*, thus measurable; and he can say that extension is qualitative and continuous without recanting the doctrine that geometrical space is homogeneous and discontinuous. Concrete extension is not the static Kantian form of matter, but matter as continuous and qualitative movement. We cannot think away matter without also thinking away concrete extension, whereas Kant grounded the apriori nature of Euclidean space mainly on his claim that we can think away matter (objects) without also thinking away space, but not vice versa. As qualitative, continuous, mobile, and concrete, this extension is beginning to look very much like Bergson's time.

Where does this leave geometrical space after DI? It has been relegated from a form of knowledge to a scheme of practical action (EC, 203), a figment constructed by the intellect for cutting up the world into discontinuous objects and events amenable to "the exigencies of action" (MM, 244) (a mobile continuity of qualities scarcely lending itself to efficient manipulation). Being purely conceptual, "space is neither inside us nor outside us" (MM, 224), whereas concrete extension is in both places. Concrete extension is not *in* abstract space, contrary to what Bergson had conceded to Kant in DI (after a richer discussion of space than Kant would have brooked). Rather, concrete extension is primary, but we project abstract geometrical space into it, thereby fragmenting and immobilizing it.

The point I have been driving at is that this timelikeness of concrete extension sustains Bergson's claim that the self–world relation is to be construed in terms of time; or that even as embracing the material world, time is autonomous of its *other*, which is space as distinct from matter. The *durée* of consciousness is not spatialized by its involvement in the material world, if it is true that the world's spatiality is time-like. But how does this temporalized space (concrete extension) differ, as it must, from the time of consciousness? What, precisely, is the nature of the time-difference between my time and the world's?

In MM, Bergson's position is inconsistent. On the one hand, his notion of matter is modelled on the body, which has a past only in the sense that it repeats the present, "acts again the past" (MM, 253). That 'again' implies no conservation of past enactments within the present enactment – no *durée*. We might say that it is a repetition so 'pure' that, were it *per impossible* to occur in human consciousness, it would fail to induce even boredom. For repetition of tokens of the same event-type bore us only because each present event-token crushes us with the weight of its conserved antecedents. But matter "does not remember its past" (MM, 250). Yet in the same book Bergson speaks of how we "suspect in nature successions much more rapid than those of our interior states"; and it is this view, according to which matter does have a minimal memory and so 'endures', that is unambiguously espoused and elaborated in the opening pages of EC (published eleven years after MM). The crucial move made here is the definition of the mind–matter difference in terms of opposed *tendencies*: "matter has a tendency to constitute isolatable systems, which can be treated geometrically [spatially]. It is even by that tendency that we will define it. But it is only a tendency. Matter does not go the whole hog" (EC, 10; see also EC, 15). If matter did go the whole hog, it would coincide with geometrical (Kantian) space, since "pure space is but the *schema* of the term at which this movement [of dis-tension of time-mind] would end up" (EC, 203; see also EC, 213). Whatever he meant to say in MM, Bergson's definitive position is that matter is the slackening tendency towards space, conceived as the ideal limit of purely quantitative repetition without memory and of mutual exteriority of parts; while mind is the opposed contracting tendency towards pure memory without repetition, maximization of self-interiority and quality, pure spirit – the sort of interpenetration of past and present exemplified by Proustian experiences of a *totum simul*. (Note that if a mind–matter dualism is to be erected on this basis it will have to be consistent with difference *of degree*.)

So, "the universe endures", it even has "a form of existence analogous to ours", "occupies a *durée* analogous to ours" (EC, 11). My *durée* and the *durées* of things 'communicate', the universe is One Great Swoosh of Interlacing *Durées*.[17] Matter endures, inasmuch as it is not *purely* quantitative repetition – less a system of rigid laws than a collection of sluggish habits (as Whitehead said). It has a rudimentary memory, its extension is ex-tension, the slackest tension of memory, "the lowest degree of mind" (MM, 250). The interiority of its past to its present is minimal, yet "not nill" (EC, 201). It is only when parts of matter are studied by physics in systems artificially closed off from the whole universe that they appear indifferent to time (EC, 10–12).

The differences between things, and between things as a whole and consciousness, are fixed by the way they realize the two tendencies in differential proportions:

> there is no single rhythm of *durée*; we can imagine many different rhythms which, slower or faster, would measure the degree of tension or of slackening of consciousness and, in so doing, would fix their respective places in the series of beings.
>
> (MM, 232)

One rhythm is slower than another if it tends less towards mutual exteriority of repetitions. Slow, therefore, means relatively contracted, 'much memory for few repetitions'; fast means slack or diluted, 'little memory for many repetitions'. Things are therefore essentially defined by their rhythms, they are time-things (as we saw) as distinct from things 'in' time. Mind differs from matter as a maximally contracted rhythm from a maximally slack one. A cow differs from a stone as a slack one from an even slacker one. The "impatience" I feel waiting for a lump of sugar to melt in a glass of water (EC, 9–10) exemplifies the general capacity of my *durée* to reveal other *durées* as non-synchronic with mine and as resisting assimilation to mine. True, the *melting process* may be speeded by applying heat – but only up to a point, which is fixed by the sugar as one way it has of affirming its obdurate *rhythm*. However we interfere with the melting process, the sugar will always 'take *its own* sweet time'.[18]

Bergson's rhythmic principle of differentiation between substances seems to do scant justice to the fact that a human *dureé* is unique in being a *subject* of experience. When I look at a cow and a stone, it is I who enjoys the privileged status of subject, not the cow or the stone. To be a subject is to differ from objects quite differently from any of the ways objects differ from each other: it is to differ from them by mode of being, not in terms of either degree or kind. Has Bergson simply failed to account for the subject? It seems to me that the nearest he comes to offering an account is in a late text (DS, 51) where he evokes the unique capacity of human *durée* reflectively to encompass and contain itself along with non-human *durées*, thereby constituting a simultaneity of fluxes prior (as he claims) to the simultaneity of instants: I can at will either live in my own *durée*, oblivious of the *durées* of the processes around me, or, by changing the direction of attention, posit my own *durée* as running off simultaneously with the co-posited external *durées*. On this evidence, then, to be a subject would be to be essentially capable of that feat of encompassing, presumably beyond the powers of non-human entities. Nice idea – but I think a relation of containment is too naturalistic to pass as an analysis of subjecthood: phenomenologically at least, the world is for-me *from beyond* me, not contained by me.

It is time we faced head-on the baffling issue of Bergson's apparent back-tracking from differences in kind (stark dualisms) towards differences of degree, his apparent re-mixing of what he had begun by unmixing. The question is: does his talk of differences of degree simply contradict his earlier insistence on differences in kind, as some commentators have supposed; or does it contribute towards a more sophisticated position in which differences in kind are somehow reconciled with differences of degree?

Referring to his genial idea that human memory produces secondary qualities by contracting the extremely rapid, almost purely quantitative repetitions of matter (MM, 203, 279), Bergson writes:

> If the humblest role of mind is to connect the successive moments of the *durée* of things, if it is in that operation that it makes contact with things and *also through it that it [mind] distinguishes itself from them*, we can conceive an infinity of degrees between matter and fully developed mind.
>
> (MM, 249; emphasis added)

But a difference of degree appears to be precisely what Bergson does not want, for he has intimated on the first page of MM that his solution will be "distinctly dualist", which means that mind and matter must differ *in kind*. Here Deleuze has come to his rescue with an ingenious reconstruction based on the notion that "*durée* is what differs from itself"; or more exactly, *durée* comprises an "infinity" of degrees of differing-from-itself.[19] We might wonder how this helps, considering that, if time differs from itself by an infinity of degrees, this is still a case of differences of degree. Deleuze's answer is that degrees of *temporal* difference structure differences in kind. Degrees of temporal difference are quite unlike spatial differences of degree. Taken 'off-the-peg', without tailoring to time, the concept of difference of degree is modelled on the spatial case: pieces of copper piping which differ in degree of length or gauge do not thereby differ in kind; by contrast, there is nothing 'mere' about differences of degree when they are specifically differences of degree *of time's self-differing*. Bergson's *durée,* according to Deleuze, is a monistic difference-substance differentiating itself into an infinity of degrees of self-differing which correspond to differences in kind. A monism of difference underlies the diversity of kinds, as well as the difference in kind between self and world.

Is this difficult interpretation sustainable? It would certainly remove the mentioned appearance of self-contradiction; and it would throw light on Bergson's otherwise obscure insistence that his dualism, though distinct, is not the "vulgar", metaphysically ultimate, dualism of Descartes (MM, 202, 249, 250). Yet it might be thought that there is something deeply unBergsonian about the notion of 'time differing from itself': do Bergson's metaphors not press in the opposite direction of time 'sticking to itself', "snowballing" (EC, 232)? Deleuze would answer that this self-adhesive property of time is indicative, not of time not differing from itself, but of the fact that its self-differing is "independent of every form of negation". But I shall have more to say about this later.

Bergson's omission of time: (i) Russell's version

Merleau-Ponty, Sartre and Russell have all of them convicted Bergson of omitting time from his theory of time (no less!). I want first to examine Russell's version of the charge since this will lead us to a consideration of the structure of *durée* and thus to the point where the French phenomenologists' version of the sin of omission takes off.

More than any other single factor it was Russell's criticisms which discredited Bergson in analytical quarters. And I suspect that of all his criticisms, this one – that the illustrious philosopher of time omits time – has done the most damage. But is it valid?

Having quoted Bergson's definition of the past as "*that which acts no longer*", Russell says:

> in this statement as indeed throughout his account of duration, Bergson is unconsciously assuming the ordinary mathematical time; without this his statements are unmeaning. What is meant by [the statement] except that the past is that of which the action is past? The words 'no longer' are words expressive of the past; to a person who did not have the ordinary notion of the past as something outside the present, these words would have no meaning. Thus his definition is circular.[20]

Let us get clear before proceeding about what Russell means by the past being outside the present. Because in general "[t]here is no logically necessary connection between events at different times", it follows that "the occurrences which are *called* knowledge of the past are logically independent of the past; they are wholly analysable into present contents". Whence it follows in turn that "there is no logical impossibility in the hypothesis that the world sprang into being five minutes ago"[21] and that we all 'remember' a history that never happened. So the past is outside the present (time is mathematical) in the sense that memory-beliefs, which are present events, entail no past events.

That Bergson's definition is circular will be granted. The damage, however, is minimal; it is just that Bergson should have known better than to try to define a primitive temporal term. A more important consideration is that the sufficient reason for the charge of circularity is just that anyone who understands 'no longer' has already understood 'the past'. Russell's specification of the past "as something outside of the present", which he presents as part of the sufficient reason, does no work at all: it seems to be there just to make the justification of the charge of circularity double up as justification of his further (and logically independent) charge that Bergson's statement is meaningless except on the assumption that the past is outside the present.

The fact that Bergson's statement fails as a definition does not of course imply that it is false as a mere statement not purporting to be a definition. But false it is, according to Russell:

> if Bergson's account were correct, the present moment would be the only one in the whole history of the world containing any activity. In

> earlier times there were other perceptions, just as actual in their day as our present perceptions. The past in its day . . . was in its intrinsic character just what the present is now. This real past, however, Bergson simply forgets; what he speaks of is the present idea of the past . . . The whole of Bergson's theory of duration and of time rests throughout on an elementary confusion between the present occurrence of a recollection and the past occurrence which is recollected . . . [W]hat Bergson gives us is an account of perception and recollection – both *present* facts – and what he believes himself to have given is an account of the relation between the present and the past. As soon as this confusion is realized, his theory of time is seen to be simply a theory which omits time altogether.[22]

Notice, first, how these remarks trade on a transparent misrepresentation of Bergson's use of the 'definite' article. When Bergson says that only the present is active he is using the article generically – saying the equivalent of '*every* present is active' (cf. 'the whale is a mammal'). Yet the 'the' of Russell's "the present moment" has the specific reading requisite to accusing Bergson of holding the preposterous opinion that *only one* present, namely the current one, is active, and that past presents were not. More important than this guying, however, is Russell's failure to see that Bergson's concept of 'remembering' is wholly different from his own. For Russell, remembering is the act of entertaining a memory-belief, something we do now and then, and which is not necessary to the existence of a consciousness. It is an explicit act performed by a conscious 'I' (what Husserl calls "an egological act"). And our present performing of the act does not entail the real pastness of what it purports to remember, whence the sceptical problem associated with this sort of memory. Bergson's memory, by contrast, is internal to the essential change of consciousness and cannot be analysed into a present epistemic act and a putatively real past external to the act. Now, Bergson is quite familiar with Russell's memory, which he distinguishes from his own in the text below (where I have italicized the reference to Russellian memory):

> [*durée*] is a memory but not *a personal memory, which latter is external to what it remembers, distinct from a past whose preservation it supposedly secures*; it is a memory internal to change itself, a memory which prolongs the before into the after and prevents them from being purely instantaneous items appearing and disappearing in a present which would be incessantly reborn.
>
> (DS, 41)

With this, Bergson roughly anticipates Husserl's distinction between *Retention* and *Wiedererinnerung*: see p. 45. His denial that the Russellian concept of memory applies to our *original* relation to the past is implicit in his doctrine, already noted, that pure memory places us *immediately* in a past it creates. This means that remembering occurs *in the past*, and is therefore *not* analysable into *present* contents (though this means that Bergson owes us an account of *mis*remembering). What Russell does not see, in caricaturing Bergson as making a hash of the empiricist theory of memory, is that, with Bergson, 'perception' and 'memory' do not refer to acts subject to pre-existent time-determinations, such that both could be said to both occur in the present: rather, they are moments in the passive and impersonal self-creation of time, which does not occur in the present, in time at all.

So, if Bergson does omit time, it is not for the reason alleged by Russell, for it is just not true that he deals with the relation between perceiving and recollecting *as distinct from* the relation between present and past. He deals with the former relation as *creative of* the latter.

It must be admitted, however, that the picture of Bergson's memory emerging here is extremely strange, perverse in the etymological sense. This memory not only takes place in the past, it is also a movement *from* the past, a movement which "prolongs the before into the after". Or even more explicitly: "The truth is that memory does not consist at all in a regression from the present to the past, but on the contrary in a progression from the past to the present" (MM, 269). And this progression is time itself, Bergsonian time creating itself. But we might allow ourselves to wonder how the past, which progresses into the present, has become past in the first place. Given Bergson's "special" sense of memory, his identification of time with memory, this progression of the past into this present implies that his theory of time takes off *after* the original self-temporalization of time has 'happened'. But the seriousness of this charge obliges us to look more closely at the structure of *durée*.

The structure of *durée*

The analytical philosopher Richard Gale once accused Bergson of subscribing to "the blob theory of time",[23] according to which time is wholly continuous; while in a later comment he says that he "does not believe that Bergson meant to assert the absurd view that time as a whole is continuous in his [non-mathematical] sense, in which case it would not be possible to discriminate between earlier and later events".[24] This change of mind epitomizes the difficulty many of Bergson's readers have experienced in reconciling the two claims he packs together under such labels as "qualitative multiplicity" and "heterogeneous continuity": (i) that time is continuous, and (ii) that, nonetheless, time is heterogeneous, thus involves succession/the discrimination of earlier from later.

In my opinion, only Deleuze has offered a *prima facie* plausible way of making sense of the conjunction of (i) and (ii), with his idea that Bergson's *durée* is structured by difference without negation. For this enables us to understand it as an evolving whole wherein 'elements' are *other* than each other without *not-being* or *excluding* each other, internally differentiated without being numerically distinct, successive without being discontinuous. The continuity in question is then not the swelling of blob-like sameness, any more than the heterogeneity is (Cartesian) discontinuity. We discriminate between earlier and later phases of *durée* in terms of a change affecting the whole qualitative multiplicity: the experience of time passing is the experience of one qualitative multiplicity *passing into* another, producing an enrichment of qualitative complexity (in a way implied by the musical example). The succession vehicled by that 'into' involves heterogeneity without discontinuity (negation). Discontinuities are produced by the needs of pragmatic action (which cuts up the environment the better to manipulate it) and by the spatial analysis applied to physical and biological systems isolated for convenience.

On the other hand, if negation is made a requirement for heterogeneity, then a present which does not exclude (negate) the past – a present interpenetrating with the past – can only be a present containing the past. Speaking on behalf of many of Bergson's critics, Sartre implicitly (and predictably) appeals to the principle 'no-difference-without-negation' to convict Bergson of obliterating the difference between the past and the present:

> this interpenetration, which is the organization of the past with the present, comes ultimately from the past itself and is only a relation of *habitation*. The past can certainly be conceived as being *in* the present, but by making it such we have deprived ourselves of every means of portraying this immanence otherwise than as that of a stone at the bottom of a river. The past can indeed haunt the present, it cannot *be* it.[25]

As against this, the Deleuzean way of making sense of Bergson has him saying (i) that the past is different from the present, and (ii) that this difference involves no kind of relation of negation between the past and the present: neither Russell's external negation (the past is outside the present) nor Sartre's ecstatically internal negation (the past is the outside *of* the present). But notice that the Deleuzean solution attempts to make qualitative difference work as a principle of temporal difference. Whether this succeeds is an issue I shall broach shortly.

More than once I have touched on the 'evolving' nature of *durée.* But just how does it evolve (change holistically)? Bergson's answer is that it evolves through "the continuous elaboration of the absolutely new" (EC, 11). There are two claims here: that the time-process is the *continuous* creation of novelty, and that the novelty created is *absolute.* I shall come back to this notion of absolute novelty later. Meanwhile a brief comparison with A.N. Whitehead will help us pin down Bergson's position on continuous creation.

According to Whitehead, time is comprised of discrete atomic durations, each characterized by internal indivisibility, which is to say that these durations are non-mathematically continuous, rather than composed of a dense continuum of events, no one of which has an immediate successor or predecessor. Each duration has an immediate successor and predecessor, from which it is separated by absolute discontinuity. Time is a process of "creative advance" in which reality becomes present in *discontinuous* bursts of creation ("sheer successions") inaugurating absolute novelty with no grounding in latency: there is a creative becoming of atomic continuities (discrete durations, "droplets" of time), but no continuity of that becoming, no smooth transition from one discrete continuity to the next. Whereas for Bergson time is precisely a *continuous* becoming of continuities, which are therefore never discrete. Time is not a series of droplets, any more than it is a swelling blob.

This continuous becoming of continuity has, for Bergson, a structure which he sometimes calls "rational evolution": it is exemplified by the free act, which

> emerges from its antecedents by a *sui generis* evolution in such a way that we rediscover in the act the antecedents which *explain* it, and that [the act] nevertheless *adds* something absolutely new, representing as it does a progress beyond them just as the fruit does beyond the blossom . . . the evolution which leads to [the free act] is a rational evolution.
>
> (MM, 207; emphasis added)

The basic thought here is this: that although no later state of affairs is deducible from an earlier state of affairs, the later state of affairs always turns out to involve absolutely created elements which *in retrospect only* appear as having been latent in the earlier state of affairs and as explaining the later state of affairs. In that sense the performance of the free act creates the antecedents which explain it.

Rational evolution is also illustrated by the way the 'family look' develops: young John's look evolves the family look, turns out to be a "coherent deformation" (Malraux) of it, yet its doing so in just the way it does was unimaginable in prospect. Sartre hits the nail on the head in likening the way Bergson's deep self is evolved by its free act to the evolution of a family resemblance. The self stands to its act as

> a father [who] begets his children in such a way that the act [of the self], without following from the essence as a strict consequence, enters into a reassuring relation with it, a family resemblance. The act . . . preserves, to be sure, an indubitable irreducibility, but we recognize ourselves in it, and it teaches us about ourselves, just as a father can recognize himself in, and be taught about himself by, the son who continues his work.[26]
>
> (EN, 81)

Bergson applies the same principle to organic life:

> this reality [life] is without doubt creative, that is to say, productive of effects through which it is expanded and goes beyond itself: these effects were therefore *not given within it in advance*, and so life could not take them as ends, *although once produced they are susceptible of a rational interpretation* . . . The future appears as expanding the present. It was therefore not contained in the present in the form of a represented end. And yet, once realized, it will explain the present as much as the present explains it, and even more.
>
> (EC, 52–53; emphasis added)

In the same vein, Bergson asks us to imagine a world in which the only colour is orange. Then orange would not already be composed of red and yellow:

> (b)ut it *will have been* composed of red and yellow when these two colours come to exist in their turn. The original orange will then be able to be envisaged from the double point of view of red and yellow.
>
> (DSMR, 313; see also PM, 18)

Or again: "once you have learnt to swim you will see that the mechanism of swimming is connected with that of walking. The former prolongs the latter, but the latter would not have initiated you into the former" (EC, 194; see also EC, 27).

Any context of rational evolution implies Bergson's analysis of possibility, which it will be worth looking at directly (see PM, 13–16 and PM, 109–116). Trivially, the fact that *Hamlet* was written demonstrates that it was logically possible when it was written. But this does not mean that before Shakespeare wrote the play it enjoyed a twilight 'possible existence' in some human or divine mind: "it is clear that a mind in which Shakespeare's *Hamlet* might have taken shape as a possibility would thereby have already created the reality; it would therefore, by definition, have been Shakespeare himself" (PM, 113). Even so, the fact that *Hamlet* had no ideal pre-existence *before Shakespeare wrote it* does not, for Bergson, mean that it cannot be said to *have been* ontologically possible before it was real. This ontological (as distinct from logical) possibility is something added to a thing once it exists, not something obtained by subtracting the thing's reality. It is *now* true that *Hamlet* was possible before Shakespeare wrote it; but all the time before he wrote it was not time during which it was possible; which means that its coming into existence retrojected a possibility it did not realize.

As we have seen, the theme common to all contexts analysable in terms of rational evolution is the retro-action of a present situation onto a past situation. Just for this reason, however, Bergson's exposition of the legitimate concept of possibility (as non-realizable) is easily confused with his diagnosis of the origin of pseudo-possibility (possibility as realizable). For in different senses, which Bergson does not trouble to distinguish, the notion of retro-action features in both. Diagnosing the origin of pseudo-possibility, he blames the tendency of the intellect to court "eternal truths" (PM, 14), in pursuit of which it effects a delusive "retrograde movement" (PM, 14) whereby what has been absolutely created is retrojected into the past and endowed with ideal pre-existence: "possibility is the mirage of the present in the past ... there, precisely, is the illusion". This train of thought wrongly suggests that Bergson denies *any* sense in which possibility precedes reality. Possibility precedes in the sense of the "retroactive effect" mentioned below:

> nothing stops us connecting the romanticism of the nineteenth century with what was already romantic in classical writers. But the romantic aspects of classicism only emerged through the retroactive effect of romanticism once it appeared. If there had not been a Rousseau, a Chateaubriand, a Victor Hugo, not only would we never have noticed, *there never would have really been,* romantic elements in the writers of the past ... Retroactively romanticism created its own foreshadowing in the past, and an explanation of itself through its antecedents.
>
> (PM, 16; emphasis original)

This retroactive effect is not the delusive retrograde movement mentioned a moment ago. Given classicism, no possibility of romanticism is co-given. But romanticism, once it exists, creates its prior possibility in classicism, which it 'rationally evolves'.

While this retroactive effect is perhaps plausible for the contexts considered, it seems hard to imagine it applying to external nature. Bergson scrambles over the difficulty with the help of a rhapsodic assimilation of the process of nature to the execution of works of art:

> I think it will become evident that, when he executes his work, the artist creates the possible at the same time as what is real. Why is it, then, that you will probably demur from saying the same thing about nature? Is the world not a work of art incomparably richer than that of the greatest artist?
>
> (PM, 113)

This is followed by some sparse and obscure remarks about how (our scientific view of?) the world involves "a constant refashioning of the past by the present, of the cause by the effect" (PM, 114). We need not pursue this – the point to note is just that Bergson means to bring external nature within the scope of his principle of rational evolution.

What is at stake in this principle is Bergson's desire to reconcile the thesis that the time-process is creative of absolute novelty with the thesis that it articulates a kind of sense, one that is determined neither mechanistically nor teleologically. What happens at an earlier time must contain no *prospective* possibility of what happens at a later time, lest what happens at a later time not be absolutely new (but a mere rearrangement of pre-existing entities). On the other hand, what happens at a later time must magically reaffirm and elaborate what happens at an earlier time, lest the time-process disintegrate into a series of unconnected droplets of creation making no sense.[27]

However, the notion of absolute novelty (to which I now return) is easily portrayed as a contradiction in terms: the absoluteness of novelty would seem to unhook it from the background of sameness against which it must stand out. A possible defence of the concept might follow Deleuze (to the rescue again!) in prioritizing novelty over sameness, heterogeneity over continuity, so that, rather than sameness featuring as the ontologically prior background against which novelty must stand out, novelty is promoted to the ground against which sameness is produced as a secondary effect (as well as the derivatives of sameness: continuity, repetition, resemblance, etc.). Novelty is thereby set up as existing absolutely, unsupported by sameness, of which it is the "groundless ground". Differences would then be what makes resemblances possible – "only differences can resemble each other";[28] whereas common sense and Platonic metaphysics opt for the fundamentalness of sameness, with its implication that only things which resemble each other can differ and that absolute difference is impossible. Turning this assumption on its head means "reversing [the] usual direction of thought" (PM, 214) and thinking *away from pure difference* down towards the lesser differences grounded in it. As John Mullarkey has pointed out: "Placing this dual option before Bergsonism one could write that novelty or heterogeneity is the ground on which continuity is built and subsequently try to justify this preference through the added weight of empirical evidence."[29] On this option, "resemblance subsists but it is produced as an external effect"[30] – perhaps the retroactive effect considered above.

Bergson's omission of time: (ii) the phenomenological version

What all the examples of rational evolution have in common is something very strange – the movement of the past into the present, or as Bergson himself says, "the uninterrupted prolongation of the past into the present which encroaches on the future" (PM, 27). If this is a form of succession, it is surely not the original form: in the original form the past starts off by being future and becomes present before becoming past; it does not start off as past and prolong itself into the future via the present. Before the past can 'progress' (if it does), the present must become past, conjointly with the future becoming present and past more-than-past. This original temporalization (the self-constitution of time), is simply elided by Bergson's rational forward march of the past, as Merleau-Ponty has observed: "Against the realism of Bergson the Kantian idea of synthesis is valid, and consciousness as the agent of this synthesis cannot be confused with any kind of thing, not even a flowing thing."[31] Later in the same text he states: "Bergson makes time out of evolved time."[32] In another work he states: "Bergson describes nothing akin to a temporalization."[33]

And as Sartre confirms: "He is right, against Descartes, in suppressing the *instant*, but Kant is right against him in affirming that there is no *given* synthesis" (EN, 181), which is precisely what the past as a starting point would be.

And recall Sartre saying that with Bergson the interpenetration of past and present "comes ultimately from the past itself": this seems to be confirmed by Bergson's musical analogy, since it is only *as re-run in memory* that we have the impression of an interpenetration of phases (later ones reaffirming earlier ones without having been predictable from them). The music we actually hear unfold is punctuated by the threat of discontinuities corresponding to the absolute beginnings and ends of physical sounds. Synthesis is the act by which that threat is overcome and the temporal unity of the music achieved.

This unity is a synthesis *performed* by consciousness itself, [34] in which consciousness is both synthesizing agent and synthesized product, both performs the synthesis and exists as it. The phenomenologists' complaint against Bergson is that he treats temporal synthesis as a given datum rather than as an enactment. A synthesis that is not performed ("constituted", as Husserl says) but *given* is something inert, thinglike, a mixture like coffee and cream; it entails the reification, thus the obliteration, of time. Or as Merleau-Ponty puts it: "If consciousness snowballs it is, like the snowball and all things, entirely in the present."[35]

The form of argument here is as follows: (i) it is essential to time to temporalize itself; (ii) things are ontologically debarred from temporalizing themselves; therefore, (iii) things, including a consciousness assimilated to things, are not intrinsically temporal, but merely *in* time. It is true that this argument relies on the weighty metaphysical assumption that things are intrinsically timeless specimens of a being-in-itself which by definition does not differ from itself, temporally or otherwise. It seems to me, however, that Deleuze inadvertently supplies the means of driving home the phenomenological charge of 'realism' without that assumption when he says that "*durée* 'tends' on its side to assume or bear all the differences in kind (since it is endowed with the power of varying qualitatively from itself)".[36] So now: if the charge still sticks, it will not be because Bergson's time fails to differ from itself (as according to the phenomenologists) but because it fails to differ from itself *in the right sort of way*. "Qualitatively" looks like the wrong sort of way. The crucial question is whether internal *qualitative* difference can come up for temporal difference (as between present and past). Is temporal difference not, like the difference between self and world, one of ontological type rather than of quality? Is quality not a category applicable only to *things*? And could time's differing qualitatively from itself do the job of 'bearing' (transcendentally conditioning) the external qualitative variations between things? Well, let's say green is a qualitative variation on red; let's grant that this variation is somehow temporally grounded; if it is grounded in time's qualitative difference from itself, then either (i) time differs qualitatively from itself in the same sort of way as green from red, in which case we have a vicious regress, or else (ii) time differs qualitatively from itself in some inscrutable and unmotivated sense of 'quality'.

The thought behind all this is that the self-differing that is the difference of the past from the present is of a deeper order than the degrees of self-differing productive of the variety of time-things differing in kind. The latter are by Bergson's own account defined in terms of degrees of contraction of the difference between past and present, which difference is not reciprocally definable in terms of degrees of contractions. We might say, following some phenomenologists, that the past is the present differing from itself in the style of voiding itself into its outside, thus doing just the opposite of fixing an essence of presentness; whereas a time-thing is a differing-from-self productive of its fixed essence, past and present are then 'non-essences' and ought not to be described as differing in nature: not because they differ in degree but because both those ways of differing are incompatible with the ecstatic way the past differs from the present. Yet Deleuze says much the same things about the difference between the past and the present as about the difference between time-things: "there is a difference of nature between the past and the present"[37] – but there is also a difference of nature between (the rhythm of) a lump of sugar and (the rhythm of) a lump of wax. "*Durée* var(ies) qualitatively from itself" – but a lump of sugar fixes its essence by doing just the same thing: "it differs in nature not only from other things but primarily and especially from itself".[38] This assimilation calls for a (problematically) qualitative account of how the qualitative difference between past and present differs from the qualitative difference between a lump of sugar and a lump of wax. Phenomenologists would object that any such account would naturalize time. It is open to anyone to remark that no account, good or bad, is in fact given by Bergson–Deleuze.

If these considerations carry weight, the position we have now reached is that qualitative difference is unfit to serve as a principle of temporal difference (taking over from negative principles, such as Sartre's "nihilation" or Russell's logical negation).

Though they both invoke the Kantian principle of synthesis against Bergson, neither Sartre nor Merleau-Ponty are rooting for the specifically Kantian version of synthesis, whereby time is paradoxically generated out of two timeless sources: on the one side, atoms of sensation (Kant's "sensory manifold") and, on the other, an extratemporal transcendental ego which binds the atoms into a unitary 'this'. The real task is to understand the multiplicity and the unity as mutually conditioning *temporal* structures, such that time "gather(s) itself together by the same act as it flows away from itself".[39] The unity, in other words, is as basic as the multiplicity, not a projection on to it. The 'I' is not above time but in it as an eye, as the fact that the time-flow constitutes itself *self-consciously*, such that there is 'someone' in time: an eye/I opened by the self-affections (what Husserl calls "passive syntheses") involved in temporalization; an eye which dominates time without originating it. Original consciousness is not, as Merleau-Ponty says, "a transcendental I freely setting before itself an in-itself multiplicity and constituting it throughout; it is an I which dominates multiplicity only *by the good graces of* time and for which freedom is a destiny". Time makes me willy-nilly master of time, while ensuring that I cannot be "the absolute author of time".[40] Now Bergson is aware that Kant's version of synthesis attempts to generate time out of non-time. What need for a transcendental self ("contentless *me*": EC, 3), he rightly asks, if there are no instantaneous states in the first place, if time is not a series of "pearls" which it would be the job of this self to thread together into a necklace (EC, 3)? True, time does not 'begin' as a multiplicity of unthreaded 'pearls', but nor does it begin as an array of threaded pearls (a given synthesis), as Bergson supposes. Time is essentially temporalization, it *is* not, but temporalizes itself."[41]

Elsewhere Bergson condemns synthesis as a ready-made device of constructivist metaphysics, producing propositions of useless generality. He points, again rightly, to the uninformativeness of defining *durée* as the synthesis of the one and the many. For space and numerous other things are also syntheses of the one and the many. And time's variety of tensions is obscured by the pat application of dialectical synthesis. The dialectical synthesis of opposites "either is or is not" (PM, 207), and "it is impossible to see how it would involve either a diversity of degrees or a variety of forms" (PM, 207). On closer inspection, however, we find him endorsing synthesis precisely where it is *given* (to intuition), as distinct from applied as a ready-made formula: "we see emerge from reality the thesis and the antithesis [and] we at the same time grasp how that thesis and antithesis are opposed *and how they are reconciled*" (PM, 198; my emphasis). But is it tenable to claim that we *see the synthesis emerge from reality*? Can it be lying there, as something *given* and waiting to be spotted by intuition? Perhaps, if time is an *object* for an intuition that is a reflection. And Bergson does say that "our intuition is reflection" (PM, 95). This thought sets us on course for Sartre's re-interpretation of the status of Bergsonian time.

Various changes may befall a duck as dozily I watch it swim in to the shore: changes in position obviously, probably also changes in the play of light on its back. It nevertheless remains the same duck from the beginning to the end of my act of seeing it: *invariantly* the same, not evolutively the same – this duck undergoes no rational evolution. In cases such as this Bergson could not plausibly invoke a pragmatic-spatializing orientation of consciousness to explain the fixity of identity. So there can be something 'solid' or invariant in time, despite Bergson.[42] In Chapter 2 we shall see Husserl make much of this invariance. The point to note here is just that these perfectly common experiences of fixed identities through time undermine Bergson's claim to be giving an exhaustive account of genuine time-experience. This is not to deprecate one temporality in the name of another, but to point to the fact that Bergson has described only one among others, and that rival non-creative varieties cannot plausibly be dismissed as spatialized. Time is not a unitary concept, we have to admit a plurality of times. That Bergsonian time is something real and important (to how we lead our lives) is, I think we must concede, as evident as that it is not the whole truth about time. The problem is to delimit the domain of its validity.

A brilliant and enormously coherent attempt at delimitation was made by Sartre with his claim that Bergson, labouring under the illusion that he was describing the "original temporality" of consciousness, was in fact dealing with a derived "psychic temporality": "[w]hat Bergson puts his finger on here is the psychic, and not consciousness conceived as For-itself" (EN, 214). Psychic temporality (Sartre's *trouvaille*) is the concrete temporality, not of original or pre-reflective consciousness, but of the thinglike states of the ego or "psyche" that is constituted by impure reflection on pre-reflective consciousness. It therefore vanishes the moment we stop reflecting: "it is to reflection that it reveals itself, and it is reflection which must constitute it" (EN, 206). The ego constituted by this impure reflection includes *inter alia* states (love, hate, envy, etc.) which are not present in the pre-reflective consciousness, whereas pure reflection preserves the 'emptiness' (according to Sartre) of pre-reflective consciousness (EN, 201, 206).

A first glimpse of the difference between original and psychic temporality may be gained from considering the intuitively palpable contrast between the consciousness I have of time passing as I read, write or speak, and the sort of temporal passage that is referred to by expressions like 'his admiration turned to envy' or 'his anger abated to indifference'. It is clear that there is a difference in kind between the sort of time-flow denoted by the verbs linking these states and the sort of time-flow in which I live as I read, write, listen to someone speak, etc. The future of which I am conscious in reading the letters 'ind. . .' as requiring to be completed to 'independent' is the original future of consciousness, whereas the future I refer to in saying 'I shall love Susan forever' is the psychic future of the reflectively-produced ego-state 'my love for Susan'. In that psychic states give themselves as permanent,[43] their future attempts to mortgage the future of the original temporality from which they are derived; and often fails: in essence, as a state, love is forever; in fact, not usually so: the input, into the state, of 'impassioned consciousnesses of Susan' tends to dry up. Psychic temporality is "an object" in the sense of an objectivized consciousness or "degraded spontaneity" (EN, 215); it "is nothing other than the unity of being of the for-itself hypostatized into the in-itself" (EN, 213).

The psyche and its temporality have supplied the subject-matter of traditional psychological novelists (Bourget, Mauriac, Proust), who analyse constituted 'states of mind', in contrast to the later so-called 'stream-of-consciousness' novelists, who try to catch the movements of consciousness in their pre-reflective immediacy (Joyce, Woolf, Sarraute). Consider the following, courtesy of Proust:

> As soon as Swann could picture [Odette] to himself without revulsion, as soon as he rediscovered the kindness of her smile, and as soon as *the desire to take her away from everyone else was no longer added to his love by his jealousy*, that love *became again* a taste for the sensations which Odette's person gave him, for the pleasure which . . . And this pleasure, different from all the others, *had ended up creating in him a need for her* which she alone could assuage by her presence or by her letters . . . Thus, *by the very chemistry of his affliction*, after having *created jealousy out of his love*, he began once more to *manufacture tenderness*, pity, for Odette.
>
> (quoted in EN, 216; emphasis original)

Sartre quotes this as a period-piece of the literary-psychological practice of misconstruing the flow of psychic temporality in terms of the mechanical causality implicit in Proust's sustained chemical metaphor (the italicized elements). As against this, he recommends a "magical" principle of cohesion between states, congruent with the inexplicable retroactive effect that secures the continuity-component of Bergson's rational evolution. I shall come back to this shortly. Here it will suffice to bear Proust's text in mind as a schematic picture of what psychic temporality looks like: in reflectively contemplating his love for Odette, Swann envisages a property of his ego, whereas his pre-reflective loving is a consciousness directed at a person. And the phases through which his love passes, whatever their principle of cohesion, are phases of psychic temporality.

Consider the following by way of further leverage on the distinction: the expressions *used to be* and *will/has become* refer to modal changes (changes of mode of being) that are germane to the temporality of original consciousness and alien to the temporality of the psyche. Past consciousness is what present consciousness used to be (and ecstatically still is); present consciousness is what past consciousness has become and what future consciousness will have been; future consciousness is what present consciousness will become (and ecstatically already is). Now, while it is true that today's good mood (a state) *succeeds* yesterday's gloom, it is not the case that today's good mood is what yesterday's gloom *has become*, or that yesterday's gloom is what today's good mood *used to be*. Whereas it *is* the case that today's *consciousness* of the smiling hillsides is what yesterday's *consciousness* of the dark and angry skies *has become*, and that the former *used to be* the latter. Or to cite Sartre's example: "(w)hile it is possible for the for-itself to be its own past, it would be absurd to require of my joy that it was the sadness which preceded it, even in the [ecstatic] mode of 'not being' it" (EN, 205–206). In other words: the moments of psychic temporality are successive without being ecstatically related, while the moments of the original temporality of consciousness are ecstatically related without being successive: it is 'all at once', not successively, that consciousness is its own past present and future;[44] it is successively, not all at once, that I am in a good mood and in a bad mood.

Because it is constituted by reflection on pre-reflective consciousness, the psyche is necessarily a temporality whose moments exist, all of them, in the past (what fits with Bergson): "The reflecting act posits a psyche endowed with three temporal dimensions, but it constitutes these three dimensions exclusively with what the reflected consciousness was" (EN, 212). Thus what in original consciousness was a protention keeping the future open-ended reappears in psychic temporality as a future that is pre-established, and that I need only wait for the psychic flow to bring down my way. The psychic future does not *become* present, it simply promises to *arrive*. By contrast, the present of original temporality "requires to be" its chosen future in the awareness that this future may not materialize (EN, 212).

Now, one reason why 'snowballing' is not a perfect metaphor is that it suggests that moments of time are whipped up into a blob in which their temporal position is melted down. Not so, according to Bergson (who denied that any metaphor could capture the whole of *durée*): every moment is "preserved at its place and date" (MM, 88; see also MM, 170). One tendency in *durée* is towards an order of mutually indifferent moments – Bergson's time "conceals under the thickness of psychic flow a succession of already constituted nows" (EN, 214), "a given succession of nows" which are "temporalized without temporalizing themselves" (EN, 213); while a countervailing and supervening tendency is towards the organization or 'snowballing' of moments: heterogeneity is prior to continuity. Bergsonian time (the psyche) thus involves:

> two contradictory modes of being, since it is *already* made and appears in the cohesive unity of an organism and since at the same time it can exist only through a succession of "nows", each one of which tends to isolate itself in self-identity. It is this that renders it rather like the magical *durée* of Bergsonism
>
> (EN, 213)

So, if my admiration is going to 'turn to envy', or my friendship for Mary 'become tinged with love', then it is necessary that the nows of psychic temporality should shake themselves out of their "tranquil indifference of juxtaposition" (EN, 212) and somehow flow into each other without shedding their dates, institute a continuity supervening upon an abiding heterogeneity. But how? By "the action at a distance of earlier on later forms" [states] (EN, 215], by "magical influence" (EN, 217):

> perpetually oscillating between the multiplicity of juxtaposition and the absolute cohesion of the ecstatic for-itself, this temporality is composed

of nows which remain at the place which is assigned to them, *but which influence each other at a distance.*

(EN, 218; my emphasis)

The distance in question is the degraded trace of the 'existence at a distance from itself' of original-ecstatic time, just as the degraded spontaneity of psychic states themselves is the impure-reflective trace of the genuine spontaneity of consciousness (EN, 217). Magical action is the principle of psychic cohesion Sartre substitutes for Proust's rational causality: "[w]e must give up reducing the irrationality of psychic causality" (EN, 217). Now, the incapacity of states to cause other states is grounded in the spentness of their spontaneity: spent spontaneity, which inheres in them as "one determination among other of a *given* [not self-creating] existent" (EN, 215). Because its spontaneity is spent (degraded), an earlier state has not the efficacy to cause a later state: yesterday's humiliation cannot be the prior cause of today's bad mood. But nor can today's bad mood be its own cause, in the style of consciousness, since its spontaneity is *qua* state just as spent as that of yesterday's humiliation. Whence "it follows [*sic*] that the earlier form has to effect from a distance the birth of a form of the same nature which is organized spontaneously as a form of flow" (EN, 215). In other words, by magical influence from a distance – without the contiguity generally deemed necessary for causation – the earlier psychic form fathers a later, more organized form in which the first rediscovers itself as an after-the-event explanatory factor, congruently with Bergson's rational evolution: "It is this spontaneity which Bergson has described in *Les Données immédiates*, without realizing that he is describing an *object* and not a consciousness."[45]

Sartre's interpretation is ingenious, seductive. Yet it falls foul of the fact that Bergson's *durée* comes into its own (is experienced at its purest) at the opposite end of consciousness from reflection, so to speak – in the life of dreams (DI, 94) or when we "let ourselves live". Reflection itself, when pure, testifies to Bergson's *durée* preceding reflection, as a familiar situation he evokes (DI, 94–95) will make clear. The church clock tolls the hour, but since 'I wasn't really listening' I did not count the number of chimes. Yet they have deposited in my consciousness a total qualitative impression, "a sort of musical phrase", which "an effort of retrospective attention" (i.e. reflection) enables me to translate into a quantity: I let chimes ring out successively in my imagination, counting them, but also attending to the qualitative impression made on my consciousness by each quantity, until I reach the impression that matches the original total impression (this works for Bergson's four chimes, but what about ten?). The point is: the certainty we experience throughout of submitting to an external constraint shows plainly that the reflection consults a *durée* which pre-existed reflection, contrary to what Sartre claims. Whatever else it is, Bergson's time is not Sartre's psychic temporality.

Conclusion

These phenomenological criticisms are important because they raise the issue of Bergson's naturalism. The key word in any discussion of naturalism is 'thinghood'.

It will have been noticed that Sartre's re-interpretation deploys a different species of thinghood from Merleau-Ponty's: Sartre's thinghood is that of non-physical non-self-creating mental objects sustained in being by a reflecting consciousness, while Merleau-Ponty's is the physical thinghood of non-self-creating matter. What these two species of thinghood have in common is *ontological* (not physical) *inertia*; what these phenomenologists' criticisms take as their common target is Bergson's description of time in a naturalist language offensive to time's dynamic essence. This charge might seem to be non-suited by Bergson's constant use of the language of creativity. But the phenomenologists' point is that this language cannot be blended with the naturalist language of thinghood – time cannot be a creative *process* ("a flowing thing"). For natural processes do not temporalize themselves any more than things do. Only non-natural consciousness temporalizes itself, therefore only consciousness is temporal (as distinct from merely *in* time). We may put the phenomenologists' position this way: while Bergson's *durée* is a substance in the technical Aristotelian sense of 'standing by itself', it is also something of a substance in the ordinary pre-technical sense of compact objects harbouring 'the darkness of matter'. Bergson's *durée* is wholly independent only of space, not also of matter and material sensations;[46] and it is this restrictedness of his version of the autonomy thesis that underlies most of the phenomenological criticism. For the phenomenological orthodoxy imposes on any theory of time the duty to conform to the basic tenets of the philosophy of consciousness as it has developed from Descartes through Kant to Husserl: to conform, that is, to the conception of time as the life of a radically autonomous, diaphanous, non-natural, transcendental intentional consciousness. While this standpoint is not unassailable, it deserves deeper consideration than Bergson gave it.[47] What is certain is that Bergson could not have adopted it consistently with carrying through his grand project of conceiving a single *durée* structure true of consciousness and the cosmos alike. It is unlikely that the cosmos respects the Kantian principle of synthesis, and it is phenomenologically self-evident that my temporally-extended consciousness of a natural process is not another natural process. One can only agree with Kolakowski's summarizing remarks:

> there is on the one hand a Bergson-Cartesian (or semi-Cartesian) . . . And on the other hand there is a Bergson-cosmologist . . . To find a consistent language which would embrace both the *cogito* and the cosmos is probably impossible.[48]

2 Time as time-consciousness
Husserl

Introduction

Husserl's great book on time[1] belongs to that squandered strain of philosophical texts bequeathed to us as desultory lecture notes or 'work in progress', not written for publication, collated by disciples, yet masquerading posthumously as coherent narrative. Its Part A was published by Heidegger in 1928, having undergone the strongly interventionist and over-synchronizing editorial services of Husserl's assistant, Edith Stein. To Part A Rudolf Boehm, editor of the standard German *Husserliana* X edition (1966), added a Part B comprised of "supplementary texts" from which Stein had incoherently excerpted parts, producing a text which jumbles incompatible positions belonging to different stages of Husserl's thought.

The position on time I shall be attributing to Husserl in this chapter is the mature one he reached much nearer the end than the beginning of the period 1893–1917 featured in the title of ZB. The textual entanglement of this position with earlier incompatible positions has meant that the dating of the "sketches" comprising ZB has proved essential to revealing the development of Husserl's thought and to identifying his mature ideas. Valuable scholarly work on the dating and arrangement of texts, initiated by Boehm, has been supplemented by Rudolf Bernet's recent proposals for slightly different groupings. To this work, as also to John Brough's,[2] I am indebted for such skill as I have acquired in the art of mentally editing out elements of sketches incompatible with the mature position.

As Brough has pointed out, it is only in the light of a correct chronology of the texts that one apparent conceptual muddle shows up as a growth of thought which is interesting to witness. The period of the lectures on time coincided partly with Husserl's slow working-out of his concept of the phenomenological reduction, which he seems to have understood initially as requiring the exclusion, from the field of phenomenological self-givenness, of the object-side of intentional acts. This immature conception of reduction explains Husserl's saying that "one cannot discover the slightest thing about objective time through phenomenological analysis" (ZB, 16), a statement belied by the many pages of ZB given over precisely to the analysis of *intentionally* objective time. From the perspective of his mature conception of reduction, objects and their time are matters for investigation by reduced consciousness: they are to be described just and only as they appear through the act of intending them, as objective-noematic senses *constituted* by time-consciousness. What is suspended, then, is only the *natural-attitude interpretation of* objective time as physical time: "through the phenomenological reduction consciousness has forfeited its insertion into cosmic time".[3] Even so, Husserl's use of 'objective time' oscillates confusingly between denoting cosmic/physical time, as in the heading of Sketch 1 "The Suspension of Objective Time" (ZB, 4), and intentionally objective time, referred to a few lines later as "the time that is posited as *objective in* an episode of time-*consciousness*" (emphasis added) and contrasted helpfully with "real (*wirklichen*) objective time" (ZB, 4).

Suspending naturalistic interpretations of objective time means abstaining from genetic questions of interest to psychologists, such as how and when a sense of time develops in infants; but also, and more relevantly, from metaphysical questions concerning the existence of a physical time and the relation of experienced time to it (ZB, 4–8, 187–189, 335–340). In a late work, however, Husserl resolves this metaphysical suspense in favour of what his critics deplore as egological idealism: "true existence, whether real or ideal, *has, then, significance only as a particular correlate of my own intentionality*".[4] Intentionally objective time is thereby promoted to absolutely objective time, to 'real time'. It is clear from the remark just quoted that Husserl's reply to the charge of idealism would be that notions of absolute (as distinct from intentional) mind-independence are without "significance". They certainly have no status as Husserlian 'senses' actualized by syntheses of acts of consciousness. In ZB, however, the framework of reduction prohibits this espousal of idealism.

The main problem up for investigation in ZB (within the restraint imposed by the reduction) is the ancient Augustinian one of how it is possible that we apprehend temporally-extended objects. Part B opens with the question: "How does the unity of a process of change that continues for an extended period of time, a unity that comes to pass or develops in succession – the unity of a melody, for example – come to be represented?" (ZB, 137). How, in other words, do we explain the possibility of the fact that we perceive a unitary melody rather than just a series of instantaneous auditory presentations? The possibility of our doing so is opaque on Augustine's premise that only a durationless now exists. The now in this sense of sheer non-extendedness is for Husserl an "ideal limit" (ZB, 40) of any imaginary process of continuous cutting-away of the past and future horizons (time-field) within which this-thing-now is intended. A clue as to how Husserl construes his problem is given by his two compound-noun coinages referring to terms of a correlation: *Zeitbewusstsein* and *Zeitobject*, respectively *time-consciousness* and *time-object*. Understanding the motivation for these coinages should give us some leverage on how Husserl sees his problem.

The concept of time-consciousness (inner time) refers to an *identity between consciousness and its time*: to the time that we *are* as conscious beings, as distinct from time as either an object of consciousness or a neutral container in which things and conscious beings would endure, preserving their identity independently of time. The concept of *time-object* (outer time) refers to an *identity between things and their time*. It registers Husserl's contention (which we have seen to be also Bergson's) that temporal extension is essentially involved in the identity of a thing; or that things are tissues or packets of time: "[e]very individual thing is necessarily temporally extended"; "[o]nly as temporally extended is the tone C a concrete individual" (ZB, 315); "[t]he individual . . . is a temporal unity" (ZB, 269). In short, an individual (non-ideal) thing is not a timeless substance possessing identity at an instant.[5] This means that the question, 'How is *a* temporally-extended thing possible?' is the same as 'How do we explain the unity of the micro-time that a thing is?' By a micro-time I mean the temporal extension across which a thing's identity is constituted, the fact that "the thing itself also contains temporal extension" (ZB, 223) *before* and as a condition of its changing in the familiar tensed style of becoming past. More of this in Chapter 5.

The difference between Husserl and Bergson over time-things is that Husserl construes time as a homogeneous form of change conditioning the possibility of there being identically-enduring objects in general, a Kantian feature of his approach which precludes it from accounting for the variety of things[6] which Bergson accommodates with his concept of differential rhythms of *durée* (see above, pp. 20–22). Husserl cannot engage with Bergson's important truism that we have to wait for the sugar to melt. With Bergson, an object's time-determination particularizes it qualitatively, as this *sort of* thing, as a "nuance" of *durée*, whereas we shall see that the time-determination of a Husserlian time-object identifies it only formally, as a bare 'this' distinguished by time-position from other 'thises'. Husserl's analysis binds a time-object's identity to its objectivity, showing little sense of what sort of object is involved. Conversely, Bergson's analysis binds a time-object to its quality or material content, ignoring the problem of its objectivity (and much else associated with the bipolar-cum-intentional analysis of experience).

We have now touched on the ascetic framework of Husserl's investigation, on his governing problem, on his interpretation of his problem in terms of how the intentionality of acts of consciousness 'constitute' temporal unities. This constitution is a difficult concept which Husserl never formally defined. We may say, safely if vaguely, that it refers to the meaning-conferring performances of consciousness through which the objects necessary to consciousness come to bear certain senses, such as 'objective': thus subjective (time-)consciousness constitutes objective time. Clearly, the objectivity of a thing is not something produced in anything like the way its material properties are, by natural processes: so constitution is a non-natural bringing-about of non-natural meaning. The notion of constitution in Husserl's time-theory only becomes graspable when we see it at work; pending which it should help to think of it as much closer to a logical relation (a sense-relation) than to any kind of causal relation.

Any version of the subjective constitution of objective time, including the one Husserl struggled with for years before rejecting it, departs from the question, 'How is it possible that I perceive temporally-extended objects (events, processes, etc.) when most of all but one of the temporal part of the objects are not present?' How can I perceive what is not 'there'? The early version of constitution Husserl conceived to solve this problem follows a perceptual model in postulating a continuum of temporally-neutral sensory contents functioning as "phantasms" or "representers" standing in for past and future sensa not available to consciousness. It further postulates a correlative continuum of pre-temporal acts of apprehending which "animate" these inert pre-temporal contents with present past and future meaning (in something like the way a signifying intention animates a physical word with a meaning). So according to this scheme of "apprehension-apprehended content", time is objectivized by and for a consciousness which is a straightforwardly non-temporal consciousness *of* time. This division between consciousness and time precludes their identity – we have not yet arrived at the hyphenated concept of time-consciousness. As from about 1909 Husserl achieves that identity by expelling all acts and contents from consciousness and relegating them to the level of constituted items on a level with temporal objects; while the residual consciousness, now contentless, is identified with the pure "flow" of time. And this pure flow is charged with the twofold task of constituting its own 'temporal' extension and that of objects.

The early scheme bristles with problems. I shall mention just three, of which Husserl was well aware. It might seem that since the apprehendings are tasked with constituting time they would themselves have to be timeless; for otherwise the problem of constituting time has simply been pushed back. But if they are timeless, our consciousness of an objective succession would not take the form of a succession of consciousnesses: it would be an instantaneous embracing of contents 'in one go', that is, as co-existent rather than successive. Moreover, it would seem that the apprehendings would have to be conscious of themselves, lest consciousness be blind to itself. Husserl never accepted that consciousness could be blind to itself. Whence follows the need for a second layer of apprehendings in which the first might be conscious, and so on down a vicious infinite regress. Finally, the pre-temporal contents postulated in the scheme reify consciousness: "we should not reify (*verdinglichen*) the structure of consciousness, we should not falsify the modifications of consciousness" (ZB, 324); "[t]he *time-continuum* is not '*something real*'. Time, as time, is nothing that endures or changes; a series of temporal differences is not again in time and includes nothing identical that extends throughout the series" (ZB, 244). With this denial that time is real Husserl is not, of course, saying that time does not exist, but rather that it does not exist in the manner of any sort of thing or occurrence. But nor is he saying that time is non-real in Kant's sense of an unexperienceable condition of possibility of experiencing empirical objects. Husserl's time-consciousness has in common with Kantian time that it is a "condition of the possibility" (ZB, 332) of experiencing objects, yet unlike Kant's time it is supposed to be capable of being brought to appearance by phenomenological reflection. Sartre took his cue from the paucity of the ontological residue to justify his pronouncement that consciousness is "a nothingness": from 'not occurrently real' he brutally infers 'anti-real'. Husserl himself understands his time-consciousness a mite more positively, as a flow of *sensing* prior to sense-*data* or contents.

The subjective constitution of intentionally objective time

Husserl is rightly impressed by the fact that we perceive 'in one go' such temporally extended wholes as whistle-blasts, melodies and utterances. When you reached the end of the sentence you have just read you were not making an effort of memory to resurrect its beginning from the past. In some sense the first word was still 'together with' the last, consistently with being temporally differentiated from it. If we were to say that the first word is 'past' when we read the last, we should need another word for the different, absent, sort of pastness that befalls it after it drops beneath the horizon of consciousness and is no longer 'reckoned with' in the task of making sense of the sentence. Before that drop, the pastness of the first word is a micro-past contained in what is commonly referred to as 'what is going on now'. When the whole reading episode falls into 'the past proper' – the past that is relatively discontinuous with what is going on now – it takes with it the continuum of micro-'pasts' essential to it as a togetherness of semantic elements. We are dealing with pastness at different levels, as pre-predicatively contributing to the constitution of an enduring episode (event, etc.), and as a predicate applying to a constituted episode. This distinction is very difficult to express owing to all the temporal words having been bagged by the predicative perspective: "the constituting can only be named in accordance with [in the language of] the constituted" (ZB, 371). Henceforth, in this chapter only I shall signal with a caret ^ uses of temporal words in the pre-predicative sense: thus 'past^'.

Part of what is meant by claiming that the time of things (objective time) is constituted by time-consciousness is that the principle of the identity of things through time resides, not in things themselves, but in our meaning-conferring consciousness of them. St Augustine rescued temporal extension by grounding it in "the distension of the soul" achieved through memory (see below p. 177). In a much more sophisticated way, dispensing with recourse to the reproductive acts of memory and expectation, Husserl too will 'expand the soul'; only, with him, consciousness, which seems not to have been invented in Augustine's day, takes the place of the soul. But

Husserl also tackles a problem which seems not to have occurred to Augustine, namely, how it is possible that the constitution by consciousness of the unity of *its own* temporal phases should condition the appearing of *objective* durations. This, in a nutshell, is the problem of the constitution of objective time in Husserl.

We understand all transcendent (intentionally external) things, events and process to endure within one universal public time. We discriminate intuitively between the duration of a transcendent thing and the duration of the perceptual act (e.g. seeing) in which it is intended, although the two durations are co-extensive. We also discriminate intuitively between consciousness of the immanent act and the act itself. In summary: I am pre-reflectively conscious that I am seeing (act) a horse jump (transcendent object), and that this perceptual event is contemporaneous with countless other events in one public spacetime. So Husserl's constitutional project must account for the internal structure of, and interrelations between:

(A) public spacetime;
(B) the time of transcendent objects (events, etc.);
(C) the time of immanent objects (acts and contents);
(D) time-consciousness.

I think the tiered structure of constitution is most perspicuous as I have just ordered it – from what is most visible and founded to what is least visible and most founding. However, since the constitution actually takes place in the reverse direction, the items above will be treated in the order D C B A.

D. The self-constitution of the unity of time-consciousness

A melody gives itself to consciousness through a series of tones, each of which is first experienced as a "primal impression". A primal impression is consciousness of a sensum as now (ZB, 326), consciousness of this-now. It is therefore a form of intentionality, unlike a Humean or Bergsonian impression. To say that this impression is intentional is to say that the impression of a sound-now is neither sound-like nor itself now.

Co-actual with every primal impression, and sharing its adequate evidence, is an automatic or passive "retention", which is also a form of intentionality. This is a consciousness of the transition of a primal impression into just-past^, coincident with the advent of a new primal impression: a consciousness of the "pushing-back" of impression A by impression B, but such that the pushing-back is at the same time a holding-on to A by B.

Also co-actual with every primal impression is a "protention" towards the future^ phases of consciousness. Substantially symmetrical descriptions apply to retentions and protentions, though they differ in that a retention intends a determinate and indefeasible pastness^, whereas what is intended by a protention may be indeterminate or, where determinate, fail to materialize.

The minimal 'unit' of time-consciousness is therefore a retentional-impressional-protentional extent.[7] Not a punctual now. Primal impression retention and protention make up a "momentary phase" (*Momentanphase*) – it is at one moment that all three are actual or 'in play together'.

That primal impression automatically retains or "holds in grip" (ZB, 118) what precedes it is a truth that Husserl has no difficulty demonstrating for his favourite example of listening to a melody. He asks: "How do we know when we come to the end that anything at all has gone before, that what is present last of all is not the entire melody?" (ZB, 139). Being aware of a tone as the *last* tone implies still having in consciousness the series of sounds of which it is the last. Without that "time-field" or "past horizon" we should not possess the distinction between a last tone in the sense of the last of a succession of tones and a last tone in the sense of the first-and-last. Symmetrically, without the protentional dimension of consciousness we should not possess the distinction between a melody being interrupted and its coming to its proper end. "If perception or phantasy or both refuse to continue the melody that has begun, how do we know that something really should follow, that something is missing from the whole of the melody?" (ZB, 139).

As our examples have surely shown, retentions are not to be confused with full-blown acts of memory. Retentions are our original relation to the past^, which they constitute; whereas rememberings are reproductive acts founded on them (ZB, 33). It is retention that provides the indubitable term of comparison enabling us to know, without consulting psychologists, that the past as we remember it is often untrue to the way it actually happened: the retained past^ belongs to the domain of cogito-certainty, whereas rememberings are haunted by the possible non-existence of their intentional objects. Retentions found rememberings (make them possible) in two ways. First, in that an act of remembering, like any other act, is a temporally-extended process whose unity is grounded in the retentionally-constituted unity of the phases of the flow of consciousness of it. Second, in that they constitute consciousness as past^, and so as a target for memory to return to. Retentions themselves do not *return* to the past^: it makes sense to speak of memory re-running events, faster or slower at will, backwards from a freely selected reproduced now into the ever-remoter past; or forwards, from a reproduced now up to the present (ZB, 47–48). None of this is true of retention, which lays down the track, so to speak, along which memory runs.

As a form of the synthesis of consciousness with itself, retentions account for (constitute) the temporal unity of consciousness, in the manner shown in Figure 2.[8] The row of *t*s represents now-times *constituted by* the primal impressions located on the parallel horizontal axis – clearly the constitutional project would not get off the ground if it helped itself to pre-established time-slots for primal impressions. An impression is individuated by the *this* it constitutes as a this-now, but it neither constitutes itself as now (as we saw) nor occurs in a pre-existent now-slot.

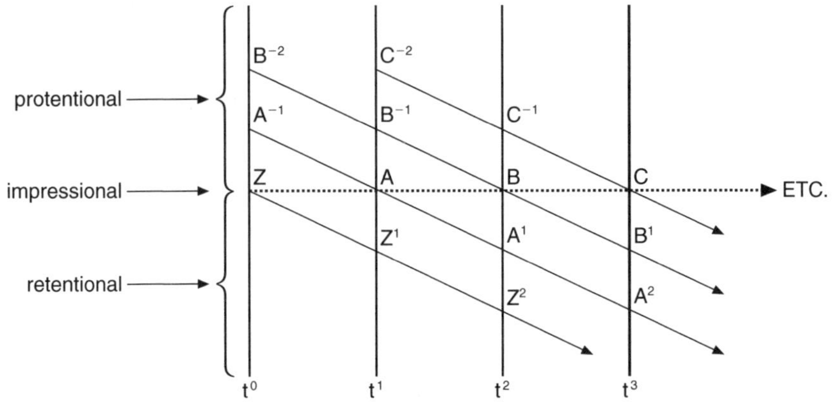

Figure 2 The temporal unity of consciousness.

It will be seen that impressions also belong to a vertical dimension. Thus considered, impression A (say) belongs together with retention Z^1 and with two protentions, B^{-1} and C^{-2}, the second intending a more distant future^ impression than the first. In that they share the same t^1, all these terms are "all at once with respect to the moment" (*momentanzugleich*). This particular 'all-at-onceness' defines a temporal phase or extent (though calling it a *phase* implies the synthesizing work of the 'transerval' retentions). A phase or extent is the 'temporal load' borne by consciousness at a time, everything 'on hand' or 'in play' in the way of impressional retentional and protentional consciousnesses. Thus at t^1 there occurs a retentive re-seeing (Z^1) of Z at a new time, and this re-seeing is co-momentary with the protentive pre-seeings B^{-1} and C^{-2}. But Z^1 is not co-momentary with Z, which preceded it. Nor is B^{-1} co-momentary with B, which will succeed it.

It will be seen further that every impression belongs, third, to a transversal dimension. Thus considered, they exist as retained (below the abscissa) by retainings co-momentary with later impressions and (above the abscissa) as protended by protentions co-momentary with earlier impressions. Considered as synthesized by retentions, the impressions on the horizontal make up a unified flux; considered in abstraction from them, they would simply substitute each other atomistically.

The mode of flow peculiar to retentional modification is the "encasing" (*Verschachtelung*) of earlier by later phases. Consider the encasing retentions comprising the retentional continuum Z^2 Z^1 Z. When B substitutes itself for A, Z^1 modifies itself into Z^2 retentive of Z^1 as itself retentive of Z. But also, since A is the impressional mate of Z^1, the retaining of Z^1 by Z^2 co-retains A and the protentions co-momentary with it, thereby interconnecting the transversal and vertical axes. As a result of the retentional syntheses, all the points on a given transversal are 'in play together', though differently from momentarily-in-play-together: they are "all at once with respect to extension" (*streckenzugleich*: ZB, 78, 375–376).[9] Thus at t^1, Z^1 is extensionally-all-at-once with Z (as well as momentarily-all-at-once with all the other data on the A-vertical), but it is *not* also extensionally-all-at-once with Z^2, despite the suggestions of the spatial figure (the *drawn* Z-line).

The effect of encasing is that the retaining at A of Z by Z^1 is retained by a second retaining (Z^2) of Z at B through Z^1. The already constituted retentional unity of the A-phase with the Z-phase is included (encased) in the unity of the B-phase with the Z-phase – and so on through ever bigger congruent triangles: at A, Z is re-seen as the self-same Z, then at B re-seen again as still the selfsame Z through the first re-seeing, now itself re-seen as the self-same. More generally, the cement of time is not retention *tout court*, but retention *of retention*. This transitivity of retention needs to be stressed in order to counter Husserl's misleading habit of speaking of retentions as if they retained only the just-past^ phase, whereas in fact the logic of his account requires that a given retention 'shake' the whole chain of retentions, including those that have elapsed far beyond intuitive grasp. Were this not so, Husserl's strategy of making retentions integral to impressions would simply substitute a multiplicity of mini-time-wholes (triangles of the same size) for the conventional multiplicity of atomic time-points, thus failing to satisfy the minimal requirement for a unitary time-flow.

Three further points should be stressed. First, what is shown in Figure 2 is only the dimension of time-consciousness responsible for its unity; the correlated structure of objective time is not represented. Second, the representation of even the structure of time-consciousness is incomplete in that it ignores the retention of protentions (which are graphed only as prospective: the minus sign denotes 'non-fulfilment'). That is, Figure 2 does not picture the fact that consciousness is retentive, not just of elapsed impressions and their retentional mates, but also of the protentional mates of elapsed impressions; that is, of having looked forward (and continued to look forward).

Third, we must beware of the spatial symbolism. Two words often used in criticism of Husserl's time-consciousness are 'linear' and 'homogeneous', both of which have exclusively spatial connotations in Bergson's vocabulary. I think the common view that Husserl's time is spatialized is a mistake induced by his diagrams. On paper, in space, the terms Z A B C of the horizontal direction exist simultaneously, whereas the real primal impressions they attempt to represent come into existence each by substituting itself for its predecessor: time cannot be *spatially* linear if it is true that "at any moment only one point on this line is real" (ZB, 235). The horizon of pastness^ at B, represented by the triangle Z B Z^2, does not exist at A, when only the horizon of pastness^ represented by Z A Z^1 exists. And so on rightwards. (I have attempted to represent this by dotting the horizontal line of generating primal impressions.) It helps to think of the diagram as being generated by a pulsating movement through which a given triangle bursts into existence, encasing the series of smaller (progressively larger) triangles. Or in Husserl's metaphor, the triangles are to be seen as a fanning-out (*Fächerung*) and not as a series of juxtaposed or superimposed fans. As to the charge that Husserl's time is spatialized because it is homogeneous, this is no better justified. It is true that his time is homogeneous in that it is throughout structured by repetitions of the same triadic relation 'retention-primal impression-protention': every part of this time is formally the same as every other. But every part *becomes* the same as every other. Husserl's time is homogeneous *growth*, whereas homogeneous space does not grow. What I contest here is Bergson's generalization from 'space is a homogeneous milieu' to '*any* homogeneous milieu is spatial'. (This is not to claim that *all* time-experience is homogenous, as Husserl implies and Bergson contests, but just that Husserl's homogeneous time is experientially real and not vitiated by spatialization.)

Retentional synthesis is analysable into (i) a synthesis of identification whereby an extent is intended (constituted) as *the very same* at a later actual impression as at an earlier one; and (ii) a supervenient synthesis of differentiation whereby the very same extent is intended as *different from* all others. (Note the non-Deleuzean priority of sameness over difference.) Temporal difference is accounted for by a form of synthesis rather than by atomistic division. Strictly speaking, extents *are* not different from each other, but rather they exist as differentiating themselves from each other – time temporalizes itself.

These forms of temporal synthesis are fundamental to conscious life, presupposed by all subsidiary forms of connectedness in experience, by every "saliency" standing out in the unitary time-flow:

> synthesis resides not only in all singular experiences of consciousness and connects not only, and incidentally, singular experiences with singular experiences. Rather, . . . the *whole life of consciousness* is synthetically unified . . . every conceivable singular experience exists as a saliency (*Abgehobenheit*) only in a total consciousness always – already presupposed as unitary [see Introduction, p. 2]. The universal cogitatum is universal life itself in its patently unending unity and wholeness . . . The basic form of this universal synthesis, which makes all other conscious syntheses possible, is the all-embracing inner time-consciousness.[10]

The *homogeneous* character of the 'growth' of Husserlian time marks it off from Bergson's time. Encasement involves no snowballing of past^ extents into anything like qualitative multiplicities. Retention "intends no new object", "yields no new now-point" (ZB, 65) – this 'memory' is not (re)creative. We first experience an impressional consciousness, then re-experience it, then re-experience it again as having been re-experienced (and so on). In neither re-experiencing is the impression re-grasped as having deviated from its original identity. At B, for example, Z^2 does not retain a qualitative multiplicity Z-*as-become*-Z^1, something different from Z in virtue of the Bergsonian principle that to be remembered is to be recreated.[11]

The effect of retention involving a synthesis of identification is that an extent now receded into pastness^ is regrasped as the same (= still the same) extent as was actual; but inasmuch as it is regrasped in the "running off mode" (*Ablaufsmodus*) 'just past^', it differentiates itself from the new actual extent appearing in the running-off mode 'now', as well as from all other extents re-appearing in a continuum of running-off modes 'more-than-just-past^' etc. The *invariant* identity of an extent emerges through (by means of as well as throughout) the *flow* of time. In that sense its invariant identity 'constitutes itself' in the flow. The dynamism of time grounds the stasis of time (though we shall see later why this version of the priority does not match the 'analytical' conception of the priority of tensed over tenseless time).

I think we are now in a position to glimpse why Husserl's notion of constitution has nothing to do with creation or production. Constitution is essentially connected with the syntheses of consciousnesses. Through these syntheses 'senses' (immanent or transcendent – immanent in the present case) are continuously actualized. 'Invariant time-relation' is such a sense, actualized through the manner of synthetic self-organization of phases of time-consciousness which I have tried to describe. It should further be remarked that the time-constituting syntheses of identification and differentiation are passive (automatic, anonymous) in that they involve no intervention of an ego which grasps itself as actively performing acts of synthesis. The syntheses in virtue of which I *perceive* one and the same building from different spatial perspectives (as I wonder whether it is the one I entered a week ago) are passive, whereas the synthesis in which I *judge* that that invariant building is identical with the one I entered a week ago is active.

One might wonder how the phenomenologist can *know* anything about this self-constituting flow of consciousness unless by means of a second-order consciousness by which the first is known – and so on *in infinitum*? The apparent problem here would be ontological as well as epistemological, since it would seem that any consciousness which might know time-consciousness would ground the latter, which would not then be self-grounding (self-constituting). Husserl meets this difficulty with the observation (which is the source of Sartre's pre-reflective cogito) that the time-flow "is so remarkably and yet intelligibly fashioned that a self-appearance of the flow necessarily exists within it, and therefore the flow itself must necessarily be apprehensible in its flowing", whence it follows that "[t]he self-appearing of the flow does not require a second flow" (ZB, 380–381). The idea is that the self-consciousness (self-appearing) of the flow is built into the retentional self-encasings constitutive of its unity: the 'eye' that espies the flow is within it, not behind it; the synthesis of identification effects a *self*-retentiveness, a folding of consciousness into itself. Within the retentional flow, "the constituting and the constituted cover each other" (ZB, 381–382), the one melts into the other (ZB, 227). This means that the flow itself emits the light by which it sees itself; rather than the light being trained on it from the outside by a second consciousness which would either be blind to itself or trigger an infinite regress in its pursuit of self-consciousness. An unconscious consciousness is an absurdity in Husserl's opinion – it would certainly scotch the reflective enterprise of bringing time to appearance. (It is this intrastructural conception of self-consciousness that finally solves Husserl's problem as to how the perception of an objective succession *can*, as it must, presuppose a succession of perceptions: the ultimate structure of that latter succession^ is such that it does not need to be unified by, and become self-aware in, a further succession of consciousness.)

Yet with this all is not made well. Let it be true that the flow sees itself without doubling itself – this self-seeing of the flow is not a seeing of the *present*, therefore not a bringing of time to original presence-for-intuition. That the present is blind to itself as present is evident from Figure 2. For consider: at A all the consciousnesses co-momentary with Z are conscious of themselves in virtue of the retention Z^1 of them. But none of the consciousnesses co-momentary with A are conscious of themselves *at* A: they are not self-conscious in virtue of Z^1, which intends only the terms on the Z-vertical. It is only at B, and in virtue of retention Z^2, that the terms on the A-vertical become self-conscious; thus belatedly, as originally past (past without having been present). As Levinas, echoing a whole battery of commentators, puts it: "the essence of all thought is perhaps the holding on to a plenitude which escapes . . . consciousness is senescence and the search for lost time".[12]

Necessarily the present cannot be present, if it is true that time-consciousness flows: the present must essentially *have been* if it is to be a phase of the *flow* of consciousness. For if the present were present, there would be a stability at the heart of time, which could not then flow. Husserl's enterprise is based on two incompatible ambitions – rendering time *present* (visible) and rendering it present *as flow*. What is compromised by this failure of the present to be present-to-consciousness is Husserl's claim that the time-consciousness of which he displays the self-constitution is original time. That the time he describes is phenomenologically real is beyond question; but it would seem to be rooted in, and lose its autonomy to, a prior original time that is 'traced' in the negative intuition of absolutely past presence (this suggestion will gather force as we proceed).

Husserl is aware of his problem: his texts provide evidence of his tendency to stifle it the moment he raises it. As below, for example:

> What about the initial phase of an experience that is in the process of becoming constituted? Does it also come to be given only on the basis of retention, and would it be 'unconscious' if no retention were to follow it? We must say in response to this question: The initial phase can become an object only *after* it has elapsed in the indicated way by means

> of retention and reflection (reproduction). But if it were intended *only* by retention, then what confers on it the label 'now' would remain incomprehensible. At most it could be distinguished negatively from that one phase which does not makes us retentionally conscious of any preceding phase; but the initial phase is in fact characterized in consciousness in a quite positive fashion. It is just nonsense to talk about an 'unconscious' content that would only subsequently become conscious. Consciousness is necessarily consciousness in each of its phases.
>
> (ZB, 119)

An objection one deems nonsensical might be worth the trouble of rebutting when it is raised by someone other than oneself, but why go to the trouble of raising a nonsensical objection *against oneself*, unless one is secretly worried that it might not be nonsensical at all? (Apart from that, the point about how the initial phase could earn the label 'now' were it not present is limp: it gets that label because it is from *it* that presence is missing.)

Another context of Husserl's subversion of his own thesis of presence is his earlier recorded characterization of the now as a durationless limit: "the now is precisely only an ideal limit, something abstract, which can be nothing by itself . . . even this ideal now is not something *toto coela* different from non-now, but is continuously mediated with it" (ZB, 41). But you can't have it both ways: either the now is a simple presence and is not mediated; or else it is mediated through retention (non-now), in which case it is not a simple presence. Husserl is faced with the following dilemma: if he gives weight to the idea that the immediate past^ is *not wholly different* from now, the danger is that this past^ will be as present as the present, in which case the difference between the two vanishes; on the other hand, if he tries to escape this by giving weight to the *difference* instead, he risks assimilating the 'little' difference between primal impression and retention to the 'big' difference between perception and reproduction (a generic term covering remembering, imaging, paying attention, reflecting, etc.). One sketch in which Husserl talks of the *perception* of retained pastness^ misleadingly suggests that he jumps on the first horn of the dilemma ('not wholly different'):

> There is no question at all here of a continuous mediation of perception with its opposite . . . if we call perception the *act in which all origin lies*, the act which *constitutes originally*, then *primary memory* [Husserl's early name for retention] is *perception*. For only in primary memory do we see the past, only in it does the past become constituted – and constituted presentatively, not re-presentatively.
>
> (ZB, 41)

But as Derrida has pointed out, Husserl is not saying here that what is retained exists now, despite its being 'perceived' and "perception [being] the act that originally constitutes the now" (ZB, 41). The non-reproductive, originally constituting character of retention is Husserl's reason for calling it a perceiving, despite the implied oddness of 'perceiving what is absent'. We can see Husserl's verbal difficulty: if he were to spend the expression 'non-perception' on the retained past^, what negative expression would be left him to designate the reproduced past, which is much more radically discontinuous with perception than retention? Derrida states:

> We may therefore surmise that if Husserl nevertheless calls [this presentation of an absence] a perception it is because he is anxious that the radical discontinuity should fall between retention and reproduction, between perception and imagination etc., not between perception and retention.[13]

That Husserl ultimately regards retained pastness^ as *absent* is confirmed by texts where, less concerned to uphold the distinction between retention and reproduction, he straightforwardly affirms that this pastness^ is "the opposite" of the now. Thus:

> if we connect the use of the word 'perception' with the differences of givenness with which temporal objects come on scene, the *antithesis* of perception is the primary memory and the primary expectation (retention and protention) that occur here; so that *perception and non-perception continuously blend into each other*.
>
> (ZB, 39; emphasis added)

But with this the cat is out of the bag: Husserl is saying here that retained pastness^ is (i) *the opposite* of presence (perception); and (ii) *within* (blended with) presence. Non-originality (absence) erodes originality (presence), the antithesis has always-already sucked the blood of the thesis. We 'begin' with the absence of a beginning, with original non-presence. "The living now . . . is always already a trace", says Derrida, for

> [a]s soon as we allow this continuity of the now with non-now, of perception with non-perception *within the domain of originarity* common to originary impresssion and retention, we receive the other into the identity-with-itself of the *Augenblick* [instant]: non-presence and non-evidence into the wink of the instant; and it closes the eye. This alterity is even the condition of presence, of presentation and therefore of *Vorstellung* in general, before all the distinctions which might be produced within it.[14]

This means that the difference between retention and reproduction cannot be the *radical* difference between presence and non-presence, but rather the non-radical difference between original and non-original non-presence.

Contrary to what Husserl claimed, representation (reproduction) is not founded on presentation but on a prior representation unsupported by presentation.

Let us see whether these potentially flashy remarks check out in relation to Husserl's theory of memory (a species of reproduction). Alterity is involved in memory, as Husserl would concede, inasmuch as the events, etc. we remember are not, as remembered, 'bodily' present, but absentively present. The issue between Husserl and Derrida is whether this absentive presence is founded on presence or on non-presence. Now, it is a general phenomenological principle with Husserl that inadequate evidence (including absentive presence) rests on adequate evidence; thus memory, specifically, rests on retention (as we have seen). Whence it follows that for Husserl the alterity involved in memory must be due to a deficiency of inner vision, since retention ensures that the whole past^ is 'there' to be remembered. The extent of retentional regression is unlimited: "[t]his process ["pushing-back"] must evidently be conceived as capable of being continued without limit, although in practice actual memory will soon fail" (ZB, 70). No remote sections of the retained past^ can in principle drop off into a no-longer-retained unconscious past.[15] The picture we are to find pleasing is one of remembering acts running down a pre-established track (the retentional continuum), in principle right down to the moment of birth. But surely a condition of remembering is precisely the opposite, namely that our original retentional nexus with the past^ has snapped, that we have *forgotten*, that there is a *de jure* limit to the retentional continuum itself. Phenomenologically, memory is the re-opening of an 'eye' that has closed, it testifies to a blind spot from which the past is brought back still drowsy with the total *non*-evidence that is forgottenness. But there is no room for forgottenness, therefore no possibility of remembering, in Husserl's picture of the whole past^ as retentionally chained back to present consciousness. His account of memory as a species of representation founded on presentation (intuitive consciousness) requires that rememberings always be founded on the retentional continuum, whereas the fact of the matter is that we remember much further back than the limit of that continuum. Memory is an absentive presence founded, not on *presence*, but on the *non-presence* that is forgottenness. So this round at least must go to Derrida.

A word by way of redressing the balance, however. Too many commentators have focused too gleefully and narrowly on the unavailability of presence in Husserl's theory, in apparent unawareness of the fact that the huge complexity of philosophical time-problems leaves vast tracts of Husserl's multi-layered analysis of time not just valid but relevant to issues in other traditions of philosophy of time, particularly the analytical. Preoccupation with presence has obscured the fact that traditional time-problems are localized within relatively narrow conceptual parameters from which the parameter of the dis-location of presence is irrelevantly remote. (My confrontations of Husserl with analytical philosophy in Part II are staged in the spirit of vindicating these remarks.) Pull the plug on presence if you will, the baby remains in the bath, with the problems of its anatomy – I mean there is no justification for treating the de-centring of the present as a pretext for ditching Husserl's problematics of time as somehow rendered *passé* by the 'overcoming of metaphysics' (metaphysics being dubiously equated with philosophies of presence). I agree with Brough that "The abiding virtue of these texts is . . . that what they have to say about time and the consciousness of time will remain vital long after our contemporary debates [about presence] have faded into history."[16] Besides, the dislocation of presence makes an opening for types of time-theories not based on presence, such as that of Levinas. I do not agree with Derrida that no theory of time not based on presence (no non-metaphysical theory) is possible, though I shall let Levinas argue the point in Chapter 3.

C. The constitution of the temporal unity of acts

We have seen how absolute or inner time-consciousness constitutes its own immediate unity as a continuum of retentions. We have not yet shown *how*, in so doing, it also constitutes the unity of the acts (perceivings, rememberings, etc.) that are the "immanent objects" of this consciousness. Much of what may be said about acts also holds of the transcendent objects intended in them, and it is in fact apropos of transcendent objects that Husserl offers his most explicit statement of the problem of the constitution of the time of objects or "objective time":

> The tone [a transcendent object] now sounds forth, and it immediately sinks into the past – it, the same tone, sinks into the past. This concerns the tone in each of its phases and therefore the whole tone as well . . . But how does it happen that in the face of the tone's sinking-back into the past we nevertheless say that a fixed position in time belongs to it, that time-points and temporal durations can be identified in repeated acts, as our analysis of reproductive consciousness has shown? The tone and every time-point in the unity of the enduring tone certainly does have its absolutely fixed position in 'objective' (even if immanent) time. Time is fixed and yet time flows. In the flowing of time, in the continuous sinking-down into the past, a non-flowing, absolutely fixed, identical objective time becomes constituted. This is the problem.
>
> (ZB, 64)

The general principle governing Husserl's solution is that "all objectivization is accomplished within time-consciousness" (ZB, 64). More specifically, the fixity of the time-relation (e.g. *earlier*) between one (immanent) object and another will somehow be constituted on the basis of (as an objectivization of) the fixity of the time-relations between extents of absolute time-consciousness. But how exactly?

So far Husserl's inner time-flow has been characterized as a purely self-concerned affair, apparently oblivious of objects. But it must somehow 'burst its banks' if it is to refer at all to objects and their objective time – if it is to constitute the time of objects (as well as its own unity). Just how it does this is described by Husserl under the heading of the "double intentionality" or two-track direction of time-consciousness. In one of its intentional strands, the flow intends itself, constitutes itself as unitary through its continuous self-retentiveness. This inner dimension of constitution (the only one so far considered) Husserl dubs "longitudinal intentionality" (*Längsintentinalität*), which runs lengthwise along the flow. Inseparably connected with this, however, is an outward-directed objectivating intentionality, dubbed "crossing intentionality" (*Querintentionalität*, ZB, 82, 300) which cuts out from the longitudinal dimension towards objects, thus establishing a correlation between consciousness as intending its own impression-centred flow-phases and as intending a succession of now-centred object-phases:

> We have a double intentionality in the stream of consciousness. Either we consider the content of the flow together with its flow-form: then we are looking at the primal-experience series, which is a series of intentional experiences, consciousness of . . . Or else we direct our gaze to the intentional unities, to what is intended as something unitary in the streaming-on of the flow: then an objectivity stands before us in objective time, the temporal field proper as opposed to the temporal field of the stream of experience.
>
> (ZB, 116)

Or again:

> in the one case [crossing intentionality] our gaze directs itself *through* the phases that 'cover' each other in the continuous progression of the flow and that function as intentionalities of the flow. But our gaze can also be aimed *at* the flow [longitudinal intentionality], at a section of the flow, at the passage of the flowing consciousness from the beginning of the tone to its end.
>
> (ZB, 80)

Let us be clear about the sense in which Husserl understands objective time to be "fixed". He does not mean that objects are temporally fixed in the sense of not passing away (as is the case on the analytical interpretation of fixed time as tenseless time). A tone B, for example, arises, runs through the micro-duration constitutive of its unity, then passes away. But what passes away is tone B's "timeless matter" (ZB, 417), and what does not pass away is its *time-position* between tone A and tone C. When tone B ends, it does not stop being later than tone A. What changes is only tone B's relation to (distance from) the actual now, not its relation to other items in the time-stream

(ZB, 64). The objectivity of an object's time consists in this fixity of its time-position, its matter being irrelevant. Or as Husserl puts it:

> the question of objective time and, above all, of how objective time-positions come about . . . is very closely connected with the question of the constitution of individual temporal objects and events . . . without clarification of the identity of temporal positions, there can be no clarification of the identity of an object in time either.
>
> (ZB, 64; see also ZB, 65, 69)

So the question of the constitution of objective time resolves itself into the question, How do the invariant time-positions of object (and object-phases) get constituted? We shall see that this invariance requires the contrast with subjective flux:

> But how, in the face of the phenomenon of the continuous change of time-consciousness, does the consciousness of objective time and, above all, the consciousness of identical positions in time and extension in time come about? The answer runs as follows: It comes about in virtue of the fact that over against the flow of the process of being pushed back in time, over against the flow of the modifications of consciousness, the object that appears pushed back remains apperceptively preserved precisely in absolute identity – indeed, the object together with the positing as 'this' that it underwent in the now-point.
>
> (ZB, 65)

Now back to crossing intentionality. Through this intentionality, object-points (minimal parts of objects) are constituted as "now-points", as just-past^ and as future,^ by impressions, retentions and protentions respectively. A now-point is an object-point that has received an initial identity from the impression that constitutes it as a 'this-now'. I believe that Husserl means us to understand that this-now gives a first individuation, one that is absolute in that it is not yet individuation by relative temporal position. The point is hard to substantiate textually (it is not supported by Husserl's frequent use of 'absolute', to mean just 'invariant'), though he does talk about now as "a constant moment of individuation in which time-position has its *origin*" (ZB, 66; emphasis added). If this is right, the initial individuation would be the bestowal of a *now* corresponding to the sense of *here* in which a man immersed in pitch darkness knows that he is *here*, absolutely here, despite not knowing where he is relative to things in the environment. When he does find out where he is in relative terms, his sense of being absolutely here is not disturbed, but confirmed. Something similar seems to be happening in the confirmatory passage from absolute temporal individuation to forms of individuation in terms of fixed relative time-position: there is an ongoing process of individuation, as we shall see.

Throughout the flow of crossing intentionalities an absolutely individuated object-point (and set of object-points i.e. whole object) is re-identified again and again as the selfsame object-point, in the sense of occupying the selfsame relative time-position. An object-point could not surface as selfsame through(out) the flux of its modes of appearance (now, just-past,^ just-just-past,^ etc.) if it changed its time-position with every re-intending, not even if it remained unchanged in all material respects. Conversely, the intentional re-identifying of a time-object-point is quite independent of whether or not it has changed materially: if Bergson is the philosopher who tries to think time without negation, Husserl is the philosopher who tries to think it without matter.

Watching this invariance of time-object-points surface through(out) the flow of primal impressions, offsetting itself against that flow, is witnessing "the wonder of time-consciousness" (ZB, 280), the working of the wonder of the constitution of objective time. But let us be clear about that 'offsetting' (contrast).

According to Husserl, objective time is *wholly* fixed; which means that its fixity does not crystallize out of any *objective* dynamism (such as the inscrutably "timeless" transience of bits of matter). It is true that the stasis of objective time ultimately rests on the subjective dynamism of time-consciousness, inasmuch as it is constituted by crossing intentionality objectivating fixities that do 'crystallize out' – out of the subjective time-flow. But objective time possesses no dynamism of its own; unlike time-consciousness, it does not constitute its own fixity through its own flux. So looking out on objective time, we experience an order of *externally constituted* fixities offsetting itself against a subjective time-flow out of which fixities *constitute themselves.* The contrast holds, then, between the different status of fixities on either side, not between fixities-without-flux on the objective side and flux without fixities on the subjective side, but between fixities-without-flux on the objective side and fixities forming themselves from flux on the subjective side.

The ongoing re-identification of time-object-points by time-consciousness (specifically by its crossing dimension) is not, however, the end of the story about the objectivization of time; for

> [w]ith the preservation of individual time-points as they sink back into the past we still do not have consciousness of a unitary, homogeneous, objective time. In the bringing-about of this consciousness, reproductive memory . . . plays an important role . . . The continuity of the temporal flow yields . . . the unity of the temporal object. It is this unity that recedes into the past ['proper']. Yet with this unity we still do not have full temporal objectivity.
>
> (ZB, 69, 109)

Why is memory necessary? Husserl's answer is complex and, as far as I can discover, comes in four parts, which I shall consider in turn.

As to the first part of the answer: suppose you remember event E. Necessarily you remember its micro-duration as comprised of its time-points, plus whatever material content that duration has. If E is a whistle blast, you may remember that a shriller phase began at t_2 (say), and that that phase still immediately followed t_1 at t_3 and at t_4. But this memorial recapitulation of the micro-time-order constitutive of "the identity that recedes into the past" does not prevent you from locating the whole whistle blast event anywhere you please in the series of whole events. Remembering it on one occasion you might, consistently with reproducing the fixed micro-order, place it between a set M of events and another set N; while on another occasion you might place it between the sets X and Y, and so on. Now an event that 'floated' in this way would not be in objective time. It is necessary to the constitution of objective time that memory should reproduce, not just a bare event with its fixed micro-order of time-points, but *also the time-horizon within which the event was originally located* (in the present example, immediately later than D and immediately earlier than F): "Recollection is not simply the being-conscious once again of the object; rather, just as the perception of a temporal object carries with it its temporal horizon, so too the recollection repeats the consciousness of this horizon" (ZB, 108). Only as thus reproduced within the same original time-horizon or "time-field" is E the *selfsame* event throughout all actual and possible re-intendings; that is, an event in objective time. I can execute two rememberings of events which, though perfectly alike, are numerically distinct; the rememberings of two perfectly alike sounds far apart in a melody, for example, are not rememberings of one and the same event. For two rememberings to be of the selfsame event it is necessary not just that the intentional object of each be perfectly alike (have the same micro-duration and content), but also that each be within the same time-horizon (ZB, 108).

That objective time is essentially connected with the repeated placing of events within the same time-horizon is brought out by Husserl's convincing contrast between rememberings and mere imaginings:

> in mere phantasy there occurs no positing of the reproduced now and no covering [identification] of this now with a past now. Remembering, on the other hand, posits what is reproduced and in this positing gives it a position in relation to the actually present now and to the sphere of the original temporal field to which the recollection itself belongs.
>
> (ZB, 51)

For an event to evidence itself as having really happened is for it to appear to recollection with this locatedness between nows invariantly later and earlier than it, as against the case with mere imaginings, which float in time.

The time-fields of memory overlap, just because the original time-fields memory reproduces did. This overlapping means, not just that we have *one* objective time rather than an objective time for every object, but also that we have the *de jure* capacity of regressing in memory from the past-horizon of a memorially focused experience into the horizonally-indicated earlier experience with its past-horizon, and so on "limitlessly" down the "chain" of ever-earlier experiences.

The second part of the answer as to why memory is necessary for the constitution of objective time is that there is an essential connection between my understanding a time-point as objectively fixed and my understanding that I may *freely* return to it again and again in memory (whereas retention is automatic):

> The possibility of identification belongs to the constituting of time: I can execute a memory (a recollection) of what has gone before again and again, continually producing 'anew' each temporal part with its filling and then grasping the same thing – the same duration with the same content, the same object – in the succession of reproductions I now have. The object is a unity of consciousness that can turn out to be the same in repeated acts (hence in temporal succession); it is something identical for intention . . . perceivable or perceivable again in any number of perceptions. I can convince myself 'at any time' of the identical 'it is'. Thus I can empirically experience a process in time for the first time, and I can experience it again and grasp its identity in repeated re-experiencings. I can return to it again and again in thought, and I can legitimate this thought by re-experiencing the process originally. And in this way objective time first becomes constituted.
>
> (ZB, 109)

We might say, briefly, that events belong to objective time insofar as they offer the permanent possibility of repetition in memory.

The third part of the answer is as follows. The constituting of objective time involves objectivization in the sense of putting of the original time-field at a reflective (memorially-reflective) *distance* from consciousness. A few desultory remarks attest to this carrying weight with Husserl: "the picture of time in the simple looking-back [retention] is unclear. In the clear reproduction I have the thing itself and the clearer the reproduction the more complete it is" (ZB, 109). To be conscious of what has elapsed only in retention would be to experience it as obsessing consciousness, as an "affection" we could only catch out of the corner of our eye [*hinsehen*] (if, indeed, we manage the spontaneity of looking)" (ZB, 47–48). The task assigned to memory here would seem to have more to do with constituting objectivity as visibility rather than as temporal invariance.

There being a fourth part to the answer would imply that one would be right to suspect that Husserl calls upon memory to constitute one side of the limitlessness of objective time. For it wouldn't do for objective time to come to an end at the limits of the original (retentional) time-field. A move in the direction of extending it is made by Husserl remarking, truly, that "the memorial time-field extends further back than the actually present field" (ZB, 125). But does this not cancel the requirement that memory be founded on retention? The memorial field cannot stretch back further than the retentional field if it is true, as Husserl also claims, that all representation must be founded on presentation. An even greater problem here is that in order to carry out the task assigned it, memory would have to extend back not just further than the original field but infinitely far back, as Husserl implies elsewhere: "if there were a limit, a now would correspond to it which nothing had preceded, and that is evidently impossible" (ZB, 70). Memory can come up constitutionally for the infinity of objective time only if, despite its mentioned *de facto* limits, it offers the *de jure* possibility of infinite regression into the past. Astonishingly, Husserl appears to resort to this *de jure* possibility: "a similar [to perceptual] continuity *without* this ideal limit is bare memory" (ZB, 40; my emphasis). But is memory not *de jure* limited by birth (or foetal existence)? No doubt Husserl would say that my birth has been bracketed, along with the rest of the empirical person. But is that not so much the worse for the method of reduction? (Further evidence of the incapacity of Husserl's approach to deal with beginnings and ends will emerge shortly.)

In short, objective time turns out to be a bigger whole than Husserl's series of expansions – from the punctual now to the original time-field and from this to the memorial field – can catch up with. (And note that further expansions would be needed if we were to insist that intersubjective historical time, involving events no living person is in a position to remember, is germane to the concept of objective time.)

There is, however, a deeper problem, one that comes into view in cases where constitution appears successful. Presumably Husserl is referring to time-concepts he espouses when he enumerates (ZB, 71–72), as examples of "apriori temporal laws", the transitivity of time-relations; their asymmetry; the necessity that, if two events are simultaneous in one now, they remain simultaneous ever after. It is a relatively minor question whether Husserl has exhibited phenomenologically all and only all the apriori time-laws there are. The deeper problem is the ambiguous status of apriori time laws within his constitutional project. How do we know they have not been tacitly presupposed by the project as criteria of the adequacy of its results? I do not think this possibility can be ruled out by remarking that no such external criteria are necessary for the reason that phenomenological evidences display by themselves the necessity of the truth of the propositions which express them. To see this, go back to the notion of 'witnessing the wonder of time-consciousness'; hold your hand on your heart and ask yourself what you wonder at: is it *simply* at seeing the necessity of the truth of certain time-propositions clarify itself (acquire transparency) at a pre-conceptual level? Or is it not *also* at seeing the process of pre-conceptual clarification *validating* precisely those concepts which were entrenched with us before we took the phenomenological turn – a feeling of homecoming? The latter as much as the former, surely. We would be a lot less impressed by a constitution that exhibited (as in theory it might) the necessity of apriori temporal concepts very different from the ones familiar to us. (Counter-intuitive philosophical claims, while rarely convincing, are thoroughly unconvincing when made in the name of intuition itself.) But this raises the suspicion that the whole undertaking of constitution must move in a self-congratulatory methodological circle, that the need for external criteria of adequacy means that Husserl has no choice but to be guided throughout by presuppositions that are supposed to have been neutralized by the reduction. The ambiguity here may be expressed in terms of a possible interference between two versions of apriori reason: phenomenological apriori reason, which is the *logos* of the *actual* experiencing of things (the 'light emitted by the fact'); and the Kantian apriori reason, which is a knowledge, possessed prior to experience, of the conditions of *possibility* of any experience (a light supplied before the fact). Are the mentioned time-laws results achieved by the first sort of reason, or are they unphenomenological elements supplied by 'Kantian reason' as a guide and goal for the project of constitution? If the former, Husserl's claim that his reductive method is presuppositionless is not undermined – but then the laws cannot do duty as external criteria of the adequacy of the constitution (which then forfeits some of its wondrousness). If the latter, they may function as external criterial of adequacy, but then the claim to presuppositionlessness is sunk. There seems to be no way in which the project can be both presuppositionless and subject to external criteria of adequacy; there seems to be no way Husserl can win.

Even if we leave all this as mere suspicion, the question now impinges as to how far Husserl's analyses are tainted by a less ambiguous reliance on unphenomenological concepts. To set us on course for an answer let us take note of another, as yet undiscussed, limitation of Figure 2: it suggests that we normally enjoy consciousness of just one (immanent act intending an) object at a time, such as the consciousness of a whistle blast prolonging itself through the series of impressions Z A B C. The usual situation, however, is that I am aware of numerous objective durations running off more or less synchronically in my sense-field, as Husserl acknowledges:

> The many primal sensations flow off, and have from the beginning at their disposal the same running-off modes, except that the series of primal sensations constitutive of the enduring immanent objects are variously prolonged, corresponding to the varying durations of the immanent objects. They do not all make use of the same formal possibilities in the same way. Immanent time is constituted as *one* for all immanent objects and processes. Correlatively, the time-consciousness of what is immanent is an all-inclusive unity.
>
> (ZB, 77)

We might liken this to the situation of two passengers getting on a train at the same station and travelling together through a number of stations. This would correspond to the fact that two co-eval primal impressions undergo conjointly the same running-off modes: "What is a being-together as an ensemble of primal sensations remains a being-together in the mode of having elapsed" (ZB, 77). As the train moves on, the two passengers change position relative to the station where they alighted, but not relative to each other. And just as one primal impression may be prolonged beyond its co-extension with another, so one passenger may get off the train at an earlier station than the other, thereby "not mak[ing] the same use of the formal possibilities".

But does this not give us as many times as there are objects with partially or wholly synchronic durations? What makes time the lengthening of *one* 'rope' rather than of a multiplicity of separate threads? We have seen how Husserl's retentional continuum accounts for the oneness of time in the different sense of the unbrokenness of any given time-*thread*. But why is time not *many* such unbroken threads? How do we account for the fact that the continuous threads are entwined into strands of *one* (ever-lengthening) rope?

Before turning to Husserl's answer, it is worth noting that it never occurs to him that the threads of time might interfere or fuse with each other, weave in and out of each other, turn back on themselves: he simply takes it for granted that our time-experience is always serial. For this he has been taken to task by recent commentators who remind us that much of our time-experience is not neatly organized into linear time-strands bound into one time-rope. More often than not it is full of twists and turns, "a hodge-podge of multiple serialities that often disrupt each other".[17] While this does not invalidate Husserl's analyses, it scotches their pretension to portray the phenomenology of serial time as the Grand Theory of Time.

But to come to Husserl's answer:

> We find many flows because many series of primal sensations begin and end. But we find a connecting form because the law of the transformation of the now into the no-longer – and, in the other direction, of the not-yet into the now – applies to each of them, but not merely to each of them taken separately; rather there exists something like a common form of the now, a universal and perfect sameness in the mode of flowing.
>
> (ZB, 77)

This universal form of flow, structured by the various running-off modes, clasps together the various time-threads into one time: "The actually present now is *one* now and constitutes *one* time-point, however many objectivities are separately constituted in it" (ZB, 71). The actually present now constitutes one time-point, as essentially *filled* by however many object-points. But nows are differentiated from another by their "ever new content" (ZB, 56), since clearly a regress is triggered by the thought that nows too are differentiated by different time-positions. The content itself is "timeless matter" upon which time is imprinted. But with this talk of form, things are beginning to look uncomfortably Kantian, blatantly unphenomenological.

One of Husserl's problems here is that the binding function assigned to the universal now-form is at odds with its individuating function. For how can a form which binds also individuate? The now-form by itself might differentiate all object-points falling under it from all object-points not falling under it; but it cannot differentiate *from each other* all object-points falling under it. It would seem that, if the form unifies, individuation cannot be fixed *solely* by the time-position it also imparts. There must be something material, non-formal, about individuation,[18] as Husserl himself casually admits: "they are distinguished from one another by their matter" (ZB, 71). Even so, he is not prepared to promote timeless matter to a sufficient principle of individuation. An object-point whose matter or content remained materially identical "still does not possess true identity with this identity of content" (ZB, 66). So this problem – essentially the ancient Aristotelian problem whether matter of form individuates – ends in a stalemate rooted in an unBergsonian separation of the form from the matter of time.

Husserl's difficulties are compounded by the fact that this now-form, which is filled with ever new content, introduces into the fluidity of time-consciousness an *unconstituted* (thinglike) element evocative of the Kantian time-frame and simply postulated to account for the unity of time-lines.[19]

Similar remarks apply to Husserl's construal of the retentional continuum (and presumably of the protentional continuum too) as a continuum of "running-of modes" (*Ablaufsmodi*), which are like a dense series of spotlights beneath which the stream of consciousness flows. Or more literally, they are "modes of appearance" under which an extent is continuously re-grasped as selfsame through different styles of givenness: selfsame under the appearance 'just-past^', 'more than just past^' etc. Under any description they are clumsy attempts to solve the problem of original temporal difference – for that is what we have run into again here. I suggested earlier, apropos of Bergson, that original temporal difference cannot be construed as qualitative difference; what is being suggested now is that it cannot be construed as due to anything like Husserl's extraneous forms or modes of appearance either. A third – at first sight promising – option is the one adopted by Sartre in his ontological reconstruction of Husserlian time-consciousness. Sartre, who, like Heidegger, laments the ontological indeterminacy of Husserlian consciousness, endows the temporal dimensions of consciousness with ontological characteristics and interprets the self-constitution of the absolute flow in terms of the transformation of these characteristics; thus, the becoming-past^ of an extent of consciousness is essentially the fact of its changing its mode-of-*being* (not of appearing) from being-for-itself to "for-itself-become-in-itself"[20] (not to be confused with being-in-itself-for-itself, which is the mode-of-being of future^ consciousness). Might these ontological differences not pass muster as original temporal differences? Perhaps, in so far as they apply to longitudinal intentionality. The problem is that they render crossing intentionality unintelligible. For objects, in becoming past, do not likewise survive under the same ontological transformation (even if they survive).

A further difficulty with Husserl's 'now' is the superfluity of incompatible characterizations offered of it. It is far from clear how now as a form of transformation (a running-off mode) coheres with either now as the correlate of an intentional impression or now as an ideal limit. And it is clear that now as this limit is quite different from the now of the following (fourth) characterization:

> it belongs to time's apriori essence that . . . the homogeneity of absolute time becomes constituted indefeasibly in the flow of modifications of the past and in the *welling-up of a now*, *of the generative time-point*, *of the source-point* of all temporal positions.
>
> (ZB, 72; emphasis added)

Without suspecting the damage it does his theory, Husserl rightly stresses the limitation imposed on the spontaneity of consciousness by this generative ("living") now:

> When something endures, then a passes over into xa′, xa′ into yx′a″, and so on. But the production for which consciousness is responsible only reaches from a to a′, from xa′ to x′a″; the a, x, y, on the other hand, is nothing produced by consciousness. It is what is primally produced – the 'new', that which has come into being alien to consciousness, that which has been received, as opposed to what is produced through consciousness's own spontaneity. The peculiarity of this spontaneity of consciousness, however, is that it creates nothing new, but only brings what has been primally generated to growth, to development.
>
> (ZB, 100)

A consciousness living in this now has no inkling of Husserl's other nows and the rational structures associated with them; which means, among other things, that this now, though a source, is not a source in the sense of the algebraic limit of a time-series – not immediately the "source of all *temporal positions*". The generative now is time as the *there Being* things and events; it conveys to us appearings which *are* as such before any positing of them as appearances.[21] Yet Husserl's whole analysis is based on the conversion, perhaps effected by reflection, of unpresentable appearings into appearances posited 'up front', as present (a conversion otherwise describable as that of the generative now into the limitative now: limits are properties of abstracted appearances, not of appearings). According to Husserl, "returning to things themselves" means "grasping them strictly as experienced". Does it? Recall the last time you heard a piece of music end – was that the end of a series of objectivating graspings of note-appearances? No, it was a withdrawal of sensory nourishment by notes that were 'always elsewhere'.

The criticism just outlined invoked Heidegger's Ontological Difference. But Husserl – or more specifically, his conception of being as being-presented – is also vulnerable to a criticism from the opposite (ontic) end of the ontological spectrum. Recall his remark: "the series of primal sensations constitutive of the enduring immanent objects are variously prolonged, corresponding to the varying durations of the immanent objects". We are tempted to add that the varying durations of the immanent objects (perceiving acts) in turn correspond to, because they are fixed by, the varying durations of events in physical time or by whatever it is, if not time, that conditions the happening and unhappening of events in "timeless matter". Yet these stoppings and startings are recalcitrant to constitution, which can at best explain only successivity, not particular successions with their particular stoppings and startings. They fall within the domain of physics, or of metaphysics as understood by Sartre: "We call metaphysics, in fact, the study of the individual processes which have given birth to *this particular* world."[22] The point is that Husserl's phenomenologically objective time is not an autonomous domain, but depends on stoppings and startings alien to consciousness and which keep poking their finger through the canvas of eidetic description. That the end of a melody is experienced *as* the end admittedly depends partly on me: there has to be a backward glance. But experiencing the end as the end depends as much on something that is not me, on something like a *(meta)physically* last note. Not that this notion helps much. But it helps a little, as much as placeholders for answers help.

It is time to ask where Husserl stands with respect to the autonomy thesis. The question turns on whether and in what sense the absolute flow of time-consciousness (comprising longitudinal and crossing intentionality) stands by itself, independently of immanent objects (acts) and of the transcendent objects intended by them. But here we must be careful to distinguish Husserl's view of the matter from the truly pertinent question of the view warranted by the evidence of the phenomena laid bare in the texts of ZB. As to the first, Brough says:

> while Husserl does distinguish the absolute flow as a dimension of consciousness from the level of constituted immanent objects, itself a dimension of consciousness, *he clearly does not think that the flow could exist* or present itself *independently* of the immanent objects it constitutes. It is in the strictest sense, distinct but inseparable from them.[23]

It is affirmed here that Husserl denies that time-consciousness is metaphysically absolute in the sense of being able to *exist* independently of immanent objects. Compare now Bernet on the same question of what Husserl thinks:

> it can scarcely be maintained that [absolute consciousness] is in an ontological sense an 'absolute' ground which 'nulla re indiget ad existendum'. It is rather the case that the phenomenology of time-consciousness compels the contrary conclusion, namely that time-constituting consciousness *cannot exist without its difference from the time constituted in it.*[24]

The quotation from Husserl "nulla re indiget ad existendum" ('requiring nothing for its existence') makes it clear that Bernet imputes to him the view, which we have just seen Brough deny he holds, that absolute consciousness possesses metaphysical autonomy. The Latin tag does not, however, directly support Bernet, since it does not come from ZB but from a later work[25] where Husserl, as everyone agrees, explicitly treats consciousness as a metaphysical absolute for reasons not germane to time.

Coming now to the textual evidence, we have seen Bernet and Brough agree that this does not justify conferring absolute metaphysical status on the time-flow. I think this reading is correct, inasmuch as the longitudinal dimension of intentionality, along which consciousness constitutes its own unity, does not *exist* apart from the crossing dimension, therefore not apart from immanent objects and the transcendent objects intended by them. But I would add that Husserl has shown the longitudinal dimension to be founding (*prior in* eidetic *meaning*) relative to the founded crossing dimension; for the unity of an immanent act is not constituted by the parts of the act itself, but results from the self-constituting unity of the phases of the *consciousness of* the act-parts. 'Consciousness of. . .' (absolute consciousness) requires a complement object, but this does not mean that it passively shadows its object. Rather, the temporal unity of the object is the shadow-side of the self-constituting unity of 'consciousness of. . .'.

B. The constitution of transcendent objects

Under the last two section headings we considered, first, how the temporal unity of consciousness constitutes itself and, second, how that constitution co-constitutes, through crossing intentionality, the temporal unity of immanent objects. This much accounts for temporally extended consciousness of temporally extended acts. In the usual case where the act has a transcendent intentionality, a transcendent time-object is also constituted as what is intended in the act; this by virtue of the correlation of intending act and intended object (strictly, object-*sense*). Further study of that act would be the study of perception (where Husserl retains the sensory contents he had expelled from the absolute self-intending time-consciousness). So there is no more to say under this rubric from a strictly temporal standpoint – unless to dispel any impression that the notion that perception 'plugs us into reality' means that Husserl's phenomenology of the temporality *of perception* has a privileged claim to be a phenomenology of original time. A phenomenology of the temporality of perception *sited in the clearing of Being*[26] could validly claim to be a phenomenology of original time – but in virtue of being so sited and not because perception is our rudimentary encounter with the world. A study of the temporality of mourning (or of other temporalities recently favoured by phenomenologists), provided it were sited in the clearing of Being, would have an equal claim to be dealing with original time. The claim is forfeited, however, by a phenomenology of the temporality of perception, such as Husserl's, where things are removed from their bashful hiding in the clearing of Being and posited up-front as exemplars of the obsessive falsehood that to be is to be-presented-to-consciousness.

A. *The constitution of universal objective spacetime*

Being-all-at-once-with-respect-to-extension is a togetherness in the sense of the *continuity* of extents. Within this togetherness, present^ past^ and future^ are 'in play together', comprising one total movement of phases. It is intuitively obvious that they are in play together, that the present impression is displacing its predecessor and, *as it does so,* being itself displaced by the future^ extent bearing down upon it. By contrast, being-all-at-once-with-respect-to-the-moment refers to togetherness at the same time-point. Thus everything which is, say, more-than-just-past^ is together in this "form-identical" (ZB, 375) way, while also being continuously (extensionally) together with everything just-past^ and present^.

Husserl needs these terms of art for the two togethernesses because the words that naturally suggest themselves, 'succession' and 'simultaneity', are for Husserl "proper" (*eigentlich*) temporal expressions and as such inapplicable to absolute consciousness (whence my careted usage of them in referring to time consciousness). Proper usage motivates Husserl's statements that (i) "[w]e can no longer speak of a time that belongs to the ultimate constituting consciousness" (ZB, 78); and (ii) "the primal impressions [of two co-perceived events] are not themselves simultaneous, and we can no more call the phases of the extensional being-all-at-once of fluxions simultaneous phases of consciousness than we can call the succession of consciousness a temporal succession" (ZB, 78). Given that 'simultaneous' properly means 'existing, momentarily or continuously, in the same now', it follows that the impressions of absolute consciousness are not simultaneous, because that consciousness is not now: now, and togetherness in the now, pertain to immanent acts and objects. A visual and an auditory experience may be simultaneous, but not the corresponding impressions. If impressions co-appeared as now, they would be positioned in time, whereas Husserl wants to treat an impression as a form of intentionality positioning something in time. In similar vein, the extensional togetherness is not, properly speaking, a *succession* of extents of consciousness. A blaze may succeed an explosion, the act of seeing a blaze succeed the act of hearing an explosion. These are what

Husserl would call 'proper successions', ones comprised of events which begin and end, come into and drop out of play. But the *consciousness* I have of the act of seeing the blaze does not in that proper sense succeed the consciousness I have of the act of hearing the explosion, because the latter consciousness is still in play when the former is actual. Our consciousness preserves itself over the rubble of acts and events: "time is the form of infinite consciousness" (ZB, 175). The one phase of consciousness *becomes* ("melts into" ZB, 227) the other; continues itself as it (whereas the act of hearing does not become the act of seeing). But this becoming does not accommodate succession in the proper sense.

However, though extensional all-at-once-ness is not succession and momentary all-at-once-ness not simultaneity, the first clearly looks much more like succession than the second, and the second much more like simultaneity than the first. It is therefore not surprising to be told that the first is "the site for the constitution of succession" and the second "the site for the constitution of simultaneity" (ZB, 375). The unity of this simultaneity and succession is universal objective spacetime, and the fact of a thing appearing within it is germane to our concept of the thing's objectivity: it would be in a defective sense of 'objective' that someone might subscribe to the proposition that a thing is in objective time and space for just as long as I see it occupy a place within my perceptual field. On the other hand, it would be to repudiate Husserl's project of the subjective constitution of objectivity to insist that a thing is in universal space quite independent of my experience of it; to insist, that is, that time-experience is an event in prior space. Whence it follows that we face

> parallel problems: the constitution of the one all-inclusive space . . . and the constitution of one all-inclusive *time* wherein the *temporality* of the thing lies, within which its duration is inserted as well as the duration of all the physical things and events that belong to the thing's environment.
>
> (ZB, 120, emphasis added)

Let us be clear, first, about this distinction between time and temporality.

Suppose that some objective duration A has the following features: it is wholly synchronic with B and C and partially synchronic with all other durations in the visual field; it involves the enduring thing in positional change at a non-uniform rate. That set of features fix duration A's specific temporality (*Zeitlichkeit*) or "time shape" (*Zeitgestalt*), "[t]he time belonging to something physical [which] is the time of that thing" (ZB, 120). We can imagine as many different temporalities as there are variations on that set of features. "And yet" – despite this variety of temporalities – "we have only one time . . . numerically one time" (ZB, 120). The unity in question here is of a higher level than the earlier discussed unity of time-lines. These time-lines are indeed time-shapes, but our earlier consideration of them pitched the problem of their unity within the limits of *my* subjective time-field. You, who are positioned elsewhere in space, experience a different set of time-shapes, thus a different temporal unity of the sense-field. Yet we both acknowledge that these two time-fields are contained within one universal *objective* time, the same for everybody. Husserl's problem is to explain how this higher-level temporal unity is constituted in such as way as to involve the unity of all space.

That space must be constituted, rather than intellectually constructed as according to some philosophers, follows from its being "co-*perceived* in every particular perception inasmuch as the perceived physical thing *appears* as lying in that space" (ZB, 120; emphasis added). The problem is: how do we explain constitutionally the fact that, though at any moment I live in a private sense-field, I also and at the same moment live in a space subsuming my visible environment and the "unseen but visible" (ZB, 123) environment outside it? Husserl's solution, in a nutshell, consists of universalizing time by "broadening" (ZB, 123) the now-form across space, so that what is not *perceived* now is "taken over into the now":

> What was earlier perceived is not only present now as something perceived earlier but is also *taken over into the now* and posited as still presently existing. Not only is that which is being perceived now, in the strict sense of now, posited as now, but also and at the same time that which was given previously. During the flow of perception proper, not only what is actually seen is posited as enduring being in the flow of its appearances, but also what has been seen. And so too with respect to the future. What is coming to be perceived in the expectation of future phases of perception proper is also posited as now: it exists now and endures and fills the same time – precisely the same thing is true of everything unseen but visible.
>
> (ZB, 123; emphasis added)

If the now were not thus broadened, what I see now would be the whole of space; contrary to fact, my perception of space would include no co-perception of a visible but unseen space; the new visual field I create by averting my gaze would be the whole of space again, differently filled at a new 'narrow' now. Whereas the fact of the matter is that I naturally suppose that the rose I glanced at on my way down my garden path is still in space by the time I reach the garden gate and am no longer seeing it. It is still enjoying the sunshine unseen, 'back there', it exists now although it is not seen now. In other words, appearances which are originally (as correlates of retentions) past are superveniently constituted as belonging to the now of actual perception, without prejudice to their retained status: the rose *as appearing to me* is no longer now, but the objective rose is now. And *mutatis mutandis* for future things: the street I protend seeing in a moment exists in the same now as the garden path I am actually seeing, without prejudice to the protended *appearing* of the street. As taken into the now, all spaces are unified into one universal space. In that sense, space derives from time.

Realists would doubtless want to challenge the philosophical import of this or any constitutional account of universal objective space or time by claiming that the sort of conscious operations Husserl describes as constituting them are 'purely subjective' episodes occurring within an unconstituted real universal spacetime. But with this we touch on the issue of how phenomenological time stands in relation to real time, a matter for Chapter 6.

3 Time beyond being

Levinas

Introducing Emmanuel Levinas

The first intimations of Levinas's seminal ethical conception of time – a time 'beyond' Heidegger's Being and Husserl's determinations of Being – appeared with the publication in 1947 of *D'Existence à l'Existant*, a svelte masterpiece of inconspicuously perfect design, written mainly in a Nazi concentration camp though reading like the harvest of an afternoon in a summer-house. Despite numerous articles published in the forties and fifties, and an important series of lectures delivered in 1946 and not published in book-form, as *Le Temps et l'Autre,* until 1979, it was only in the wake of *Totalité et Infini* (1961) that Levinas, hitherto known only as an interpreter of Husserl, came to the fore in France as an original philosopher in his own right. *Totalité et Infini*, generally regarded as his main work, was followed in 1974 by *Autrement que l'etre ou au delà de l'essence,* his second and last 'big book'.[1] An important new theme here is the time of 'Saying' as interrupting or undoing of the synchronic time of the Said, an idea interestingly applied to explain the persistent recurrence of scepticism despite conclusive refutations. Apart from that, Levinas returns to his old themes in a less irenic tone and in the more convoluted style that has earned him the reputation of being a difficult writer. He has the strangest ways of saying the strangest things. His philosophical style, which merits separate study, forces French to work in ways in which nobody had felt the need to make it work before. This may or may not be justified by his originality, which is indisputable; reading Levinas, one soon discovers oneself in the presence of a mind inoculated against the malignancies of the intellectual market-place.

De L'Existence à l'Existant was doubtless saved from oblivion by the impact made by the big books – it had to wait more than thirty years for its first reprint, in 1978. Yet it offers a radically new conception of first philosophy located "hither side of"[2] both Husserl's theoretic (object-knowing) intentional consciousness and the pragmatic attitude of Heidegger's Dasein, which was supposed be as far hither side of the theoretic attitude as one could get. "Being-in-the-world, as they call it nowadays, is an existence with concepts"[3] is a statement that must have raised eyebrows in the heyday of existentialism. Although points of contrast with Heidegger will inevitably surface in what follows, it cannot be part of my brief here to assess the overall validity of Levinas's interpretation of Heidegger, which has been said with justification to be short of compelling in every point. Some of Levinas's remarks on Heidegger are demonstrably wide of the mark, such as that Heidegger's Being is the old-fashioned "anonymous being in general" (TI, 17), the highest genus.[4] There is also a problem with Levinas's claim (explicit on pp. 15–16 of TI) that Heidegger's subordination of beings to Being entails the inclusion of the ethical relation within "a relation of knowledge" of finite beings, ultimately within the understanding-of-Being presupposed by knowledge of beings. The Other, according to Levinas's Heidegger, could only be encountered within the horizon of an all-encompassing Being/Totality destructive of his or her all-transcending infinitude.[5] But consider this: having drawn attention to the capacity of boredom to disclose "what-is-as-a-whole", Heidegger adds, in the sealed mystery of a one-sentence paragraph: "Another possibility of such disclosure is concealed in the joy we feel in the presence of the existence – not in the mere person – of someone we love."[6] Does this joy not 'thank Being' for having gifted a person who, though standing within the clearing (context) of Being, nevertheless stands out in his or her infinitude? *That there is* the loved person is germane to his or her ethical value.

It is a pity that Levinas does not give context the attention to which it is entitled by his claim to be doing philosophy from the standpoint of "the possibility of a *meaning without context*" (TI, xii). Though meaning may be originally pre-linguistic and contextless, inasmuch as born in the contextless encounter with the Other, it is still something that comes to depend on context, pre-linguistic and linguistic. Having supplied rudiments pointing upwards towards a general philosophy of meaning crucially involving ethics, Levinas chooses to remain at the ground level of contextless ethical meaning. Yet the suggestion of the last paragraph was that this may be a loss for ethics itself, not just for the philosophy of meaning.

Be this as it may, contextlessness is the key to his philosophy. The contextlessness of the ethical encounter with the Other grounds his conception of time as a multiplicity of horizonless instants (a multiplicity that is not textured into a succession); there is time because there are other people, and the structure of time is essentially that of the ethical relation. To elucidate this is the purpose of the first section of this chapter, while the second section will argue that for Levinas the time-relation required by objective knowledge is grounded in the time of the ethical relation.

It will become apparent that I dissent from the view that the unity of Levinas's philosophy is broken by a 'turn-around' (*Kehre* of Heideggerian resonance) from ontology to beyond-ontology. In all his works Levinas is doing philosophy beyond (or before) Being; and in all of them he uses the language of ontology in order to forestall psychological misreadings. First, he uses that language fairly straightly, thus inappropriately to his sub-ontological themes; then, after TI, he bends and twists it, tries to get it to "unsay itself". From start to finish this linguistic makeshifting serves the expression of philosophical themes beyond ontology – it records no *shift* from ontology to beyond ontology.

The Other as the source of time

Before time proper

I shall be taking the mentioned early works seriously, rather than as superseded drafts of the later work, because I think they give a clearer overview than do the big books of the general scheme of Levinas's philosophy of time. This is partly because they are directly about time, as the title *Le Temps et l'Autre* announces, and as is attested by Levinas's statement that EE "is governed by the theme of time" (EE, 147).

With the early Heidegger, the questioning of Being runs into the unsurmountable problem posed by Being's status as the ultimate, and unencompassable horizon from which light penetrates back through the more accessible phenomenological horizons, illuminating them with senses of Being. But at least Being is light (before it turns nasty in the later Heidegger), whereas, for Levinas, Being is darkness and the domain of intelligibility ('light)' lies elsewhere – in "the light of the [Other's] face" (TI, 125). Standing in this light means, for Levinas, standing in the "metaphysical relation" to "the Platonic Good-beyond-Being" (EE, 13).

To give us a feel for the darkness of his Being, Levinas invites us to imagine the forms of all determinate entities obliterated (EE, 93; TA, 25). What would remain, indestructibly, would not be a *nihil absolutum*, or the inert paste Sartre calls being-in-itself, but the "*there is*" (*il y a*), the absence of forms as an unremitting senseless presence: "the murmur of silence" (TA, 26), the anonymous impingement of an indeterminate evil Be-ing, "an existing (*exister*) that occurs without us, without a subject, an existing without existents" (TA, 24).[7]

Insomnia is the situation *par excellence* where there is no need to imagine the obliteration of the forms of all things in order to get the feel of the formless insistence of the *there is*. The insomniac is enveloped by the *there is*, whose element is experienced as night. The distinction between the self and world is dissolved into an anonymous vigil "without beginnings or end" (TA, 27). In the subject-dominated time described by Husserl, events begin, elapse, end; there are enduring identities, degrees of pastness; whereas the time of the *there is* is defined by the absence of all that. "Time departs from nowhere, nothing distances itself or fades off" (TA, 27).

Time? But how can this be time, given that "we can also characterize this 'existing' by the notion of eternity", since "the existing without existent is without point of departure" (TA, 28)? It would seem that it cannot be properly called either time or timelessness. Of timelessness proper no insomniac would complain, "Is this never going to *end*?" (TA, 27; emphasis added), while wakefulness in serial time would be punctuated by beginnings and ends. The *there is* is not even a pulsating instant: "this return of presence within absence does not happen in instants . . . rhythm is lacking from the *there is*" (EE, 11). What is going on here? Our suspicion that we are being taken on some adventure starting with some quasi-time before time proper, is confirmed on reaching the heading of the third sub-section of the final section of EE: "Towards time" (p. 147).

Towards time? Is time not the skeleton of human existence? In announcing a progression towards it Levinas is not embarking on some old-style project of deducing or constructing time. Rather, he is taking the *there is* as the starting-point of an "ontological adventure" (EE, 48, 52, 172) – which is really a sub-ontological adventure – towards a time 'beyond' beings and horrendous Being; beyond the times of Husserl and Heidegger; towards a liberating time opened up by the ethical relation to the Other; a time uniquely deserving the epithet 'time proper'. In this section we shall follow that adventure (renamed "the production of the infinite" in TI).

Clearly we cannot save ourselves from the *there is* by 'getting a grip of ourselves' in an act of will, given that the self is obliterated in the experience of the *there is*. The popular remedy nevertheless alludes nicely to the fact that being a subject involves *having* a grip. But *getting* a grip could only occur courtesy of a pre-volitional 'event' of Being whereby "there emerges within impersonal being, as if by an effect of hypostasis, a being, a subject, an *existant*" (EE, 18). What has to happen is that "in the pure verb of being, in *being in general*, an *existant* posits itself, a substantive which acquires mastery over being" (EE, 16). This *existant* is the "ontological transmutation" (EE, 125–126) in which the verbal *there is* ruptures itself from itself to posit itself as a 'substantive' subject or self, imposing a stop on, thus gaining command over, "the [absolute] past from which it comes" (EE, 169). As such, the present/self is "master of its time" (EE, 169), "subject of its becoming" (EE, 168). The self is the ontological event in which Being 'nominalizes' its insistent verbality, so that the self dominates its episodes in something like the way a grammatical subject governs its predicate. Nothing images this hypostasis better than the relationship in Rodin's sculptures between the human figure and the base, which is essential to the figure's 'stance' (hypostasis) rather than serving as a conventional support (EE, 124).

Hypostasis, then, is the event of the emergence of the *existant* or subject. As a beginning, the subject is without duration: its enduring would mean its perpetuating what preceded it, its "benefit(ing) from an [Husserlian] inheritance" (TA, 33; see also EE, 125). This beginning, which Levinas also calls "the present", is therefore not a moment or stretch of *durée*:

> in positing the hypostasis we are not yet introducing time into being. In giving ourselves the present we are not giving ourselves a length of time taken out from within a linear series of duration, nor a point in that

> series. We are not dealing with a present cut out from an already constituted time, with an element of time.
>
> (TA, 32)

This difficult notion of a non-serial present will exercise us for a while yet.

A strange present, one that "comes from the past" (EE, 169). Does the past not 'come from' a present? The relative (serial) past comes from a present but Levinas's present comes from the past in that it is born *having been*, in an absolute past, the *there is* which it holds at bay. The past from which this present comes is therefore not a dimension of a constituted subject's self-interiority, not the past as a dimension of a preserving *durée.* The present of hypostasis "has a history, but it is not the history" (TA, 32), resembling in this a reformed murderer who has a criminal past without *being* it (ecstatically). The past of hypostasis is what has been disowned or put out of play to leave the field free for the game (ontological adventure) of subjecthood to begin.

An *a contrario* confirmation of the effort demanded by this assumption of subjecthood is supplied by the temporality of listening to music, where we are exempted from the necessity of accomplishing the present. Listening to music is universally acknowledged to be restful (save when we attend to technical detail). We take up no 'position', we are 'carried away', fuse with something which is never *here*. What is never *here* is the instant or present: "the instants of the melody only exist in so far as they sacrifice themselves to *durée,* which, in a melody, is essentially continuity . . . there are no instants in a melody" (EE, 46). It is only false notes (*non*-music) that stand out as present. Musical temporality is continuous, whereas "the time of effort is made entirely of stops" (EE, 48), of non-emergences into the future.

What kind of time is it that disowns the past and stops short of a future? We encounter here, at the level of hypostasis, the same sort of problem of a 'quasi time' as we ran into at the level of the *there is*. In sorting out these (quasi-)times, we get scant help from Levinas's chaotic use of the word "temps", to refer to: (i) the 'time' of the *there is* ("time departs from nowhere"); (ii) the 'time' of hypostasis ("time as the pure event of hypostasis" (TA, 34), as an "ontological schema" (TA, 32)); (iii) "hypostatized time" (TA, 34, 38; EE, 125), which is any time produced by the subject forgetting its verbal basis and setting itself up as self-dependent; (iv) the "time of economy", of which more anon; and (v) the 'time proper' opened up by the Other and completing the ontological adventure.

The difficulties for reading entrained by this multiplication of 'times' are compounded by Levinas's dubious decisions as to which of them can be said 'to be'. Thus, for example, "the 'I' is not initially an existent . . . it strictly speaking does not exist" (TA, 33). But because it is a Being-event, it cannot be said not to be either: it is "beyond the categories of being *and nothingness*" (TA, 33; emphasis added) which apply to ordinary things and events but not the self accomplished by hypostasis. Levinas misrepresents the problem when he blames it on hypostasis being an event rather than a thing – "[the present] is an event, not yet something, it does not exist" (TA, 32) – as if the ordinary empirical 'is' were not applicable to events as much as to things (analytical philosophers who prefer event-ontologies to substance-ontologies have no exacerbated problem with 'is'). What causes the problem is rather that the event of hypostasis is *no ordinary* event. The problem with 'is' recurs at the next time-level up, with hypostatized time where, Levinas thinks, there is no problem about applying 'is'. Hypostatized time is "time as an existent" (TA, 33), "a time which is" (TA, 34). But surely there is a problem. Though he believes that all times on offer in the history of philosophy are hypostatized (TA, 43), he has in mind especially the times of Husserl and Heidegger – both of whom emphatically denied that time 'is'. The plethora of ways in which times fail to be defeats anything Levinas could say or do with 'is' or 'is not'.

As I indicated, the essential fact about Levinas's present is that it is not an element of serial time. On this he insists: "before being in relation to instants which precede or follow it", the instant is "pre-eminently the accomplishing of [the self's] existence . . . a relation, a conquest, without that relation referring to any sort of future or past, to any sort of thing" (EE, 130). "Time does not bring about a succession of moments" (TI, 260). In the present instant we have "an event of the instant which does not consist in its relation with other instants" (EE, 129). "Duration does not affect the contact with being accomplished by the instant" (EE, 131). We are dealing with a distinction between the instant "in its production" (EE, 125) and the instant as produced, as serial, as hypostatized as distinct from hypostatic; between the instant as the event of positioning whereby a subject emerges and as a position in a time-series belonging to a constituted subject. Levinas's present is dialectically prior to the series of instants, which is the time-form of worldly things and events: "[t]he relation to a world is not synonymous with existence. The latter is anterior to the world" (EE, 26; see also EE, 173) – man is not *fundamentally* a being-in-the-world after all.

Thus conceived, as pre-serial, the instant is not the paltry thing to which it has been reduced by time-theories that located it within the series and took duration as the measure of being. Within the series, the instant is without duration and therefore without being; outside the series it enjoys "the fullness of a contact with being" (EE, 132). In defending the ontological dignity of the instant, Levinas sees himself as going against the grain of "the whole history of the philosophy of time" (EE, 126), which has always denied fullness of being, in the first place to time, and *a fortiori* to the instant. Traditionally, forms of existence have been hierarchically ordered in accordance with how much being they possess, as measured by how complete they are. Thus that which is in time, being incomplete, has less being (thus less worth) than the Platonic Being that is eternally complete and invulnerable to time. While to be in time at all is to be degraded, some things are more degraded than others, depending on how they are in time. Something which endures everlastingly has more being than something with finite duration; so much so that everlasting existence simulates eternity (or at least this is how Levinas reads Plato's "time is the moving image of eternity"): though the durationless instant does not by itself 'bite on' being, the endless succession of instants amounts to a simulation of timeless Being: "The extension of time appears as the extension of existence itself. Everlastingness is the superior form of existence [in time]" (EE, 127).[8] At the bottom of the scale comes the traditional instant, whose being as measured by temporal extension is zero:

> the instant has no duration . . . in its contact with being is announced its detachment from being . . . the fundamental consideration which allows this denigration of the instant has to do with the fact that, by itself, the instant has no magnitude, no duration, is not duration.
>
> (EE, 132, 126)

"Modern philosophy", which Levinas appears to date from Bergson, broke with tradition to the extent of preferring time over eternity. It nevertheless perpetuates, according to Levinas, the age-old tradition of belittling the instant, inasmuch as it privileges "the dialectic of time" over the "dialectic of the instant", failing to see that the instant possesses its "own dialectic" and "show(s) indifference to the series" (EE, 126–128). The modern vindication of time has rested on the demonstration that the past and future are 'not nothing', that time is minimally a *durée*, but this has not benefited the instant, which has been knocked into whatever shape (an ideal limit with Husserl, a snapshot of flux with Bergson) suits a particular conception of time as extension (serial or non-serial). Though philosophers no longer denigrate the instant in the name of eternity, they continue to denigrate it in the name of time. Or so Levinas thinks.

One way of strengthening our leverage on Levinas's concept of hypostasis (present/self) would be to ask whether it goes on all the time, or is a once-and-for-all accomplishment (of selfhood). The answer is 'neither'. We may indeed *represent* hypostasis as going on 'all the time', that is, at each instant of serial time: "Considered in the perspective of economic life, where instants are equivalent and compensate each other [the event of hypostasis] pertains to all moments" (EE, 26–27). If hypostasis could 'speak for itself' as it goes on, instead of being spoken for in the falsifying language of serial time with its "equivalent" (EE, 129) instants, all it could say is that it goes on as a stagnating sameness. Not in *successive* bursts.

The existentialism of Heidegger and Sartre underwrites the design and organization of work in our society, where the effort and pain of today are compensated for by the "wages" of tomorrow: we struggle *for the future*, for self-preservation, devaluing the present to an instant among others in a time-order where all are "equivalent" (EE, 179). This social time, which

Levinas variously calls "the time of the world", "the time of economy" and "the time of compensation" (EE, 26–27, 29, 154–158, 155) has no instant that is present in his sense of struggling *for itself*, rather than for a future beyond itself. However:

> Taken at the level of the time of economy where [the relation of the *existant* to its existing] is customarily considered, that relation *appears* as the struggle for a future, as concernful involvement that human being has in its duration and conservation.
>
> (EE, 29; emphasis added)

But this appearance is false of the original present (Levinas's), which is not the self's existential struggle for its future but the prior anonymous struggle for the self itself, for humanhood – "the conquest of human being which recommences perpetually *as if* it took place in Cartesian time with its discrete instants" (EE, 27; emphasis added). What the perspective of economic time covers up is that "the instants of time do not exist on the basis of an infinite series *in which they appear*, but . . . they can also be on the basis of themselves" (EE, 169; emphasis added).

Having no present, economic time is "exterior to the subject" (EE, 157), anonymous public time. It is obtained by our habit of "tak(ing) the instant anywhere in 'the space of time'" (EE, 129), whereby the defining centrality of the present is read out: "in abstract time there is an order of instants, but there is no central instant, there is not that instant that is *par excellence* the present" (EE, 129–130). On this austere definition, abstract times are produced the moment the present instant is in any way connected with other instants: to construct abstract time we need not go so far as to connect instants in a specifically spatializing way. So although Levinas "associates [him]self with the criticism levelled since Bergson against the confusion of abstract time and concrete time" (EE, 130), this does not mean that he endorses Bergson's *durée* as truly concrete time. For concrete time, according to Levinas, has a subject, whereas we saw (p. 21) that a subject is hard to find for Bergson's *durée.*

None of this makes sense if we lose sight of the special meaning 'central instant' has for Levinas. Suppose I say that a certain event took place ten instants ago. Inasmuch as 'ago' is a tensed term, I have located the event in a tensed time-series which, surely, *does* have a central instant, namely the instant we call 'now'. And so it would appear that I have *not* "taken the present instant anywhere in the space of time". I could have located the event in the tenseless series (the B-series: see pp. 140–141), in which case its time-position is fixed without reference to a central instant. But in locating it in the tensed series I link it to what is *ordinarily* understood by 'the central instant'. So what is it about Levinas's concept of central instant that entails that the now fails to qualify as central? Well, now is privileged, in that it is our perspective on the past and future. But *it is also a member of the series*, it comes after some nows and before others; its central status as a perspective is supplementary to its non-central status as one term among others of the series. Whereas Levinas's present is central in that it is an absolute beginning that cannot be put into relation with other serial instants by means of the concept of ago. His present is a power-base rather than a perspective, windowless.

Levinas's present/self is not just a beginning, but a beginning that "comes from itself". That the notion of something coming from itself flouts the law of contradiction he readily acknowledges (EE, 130–131). But he offers a metaphor which it is worth trying to develop. Of the beginning accomplished in hypostasis he says, "its point of departure is contained in its point of arrival, like a return shock" (EE, 131). Imagine you hear the hoot of an owl, just that. If you experience an end here, it is only as the unrepresentable limit dividing the hoot from whatever events immediately follow. Suppose instead that you hear the hoot and its perfect echo. It can happen that the echo stuns the hoot: the echo does not follow the hoot, but rather the hoot *as stopped* is present in the echo. The echo effect is the positive stoppedness of the hoot which caused it, the dis-empowerment of the hoot's causal efficacy: "the evanescence of the instant constitutes its very presence" (EE, 132). This leaves nothing external to the echo (self) from which it might come: it is left to be by itself (the lonesomeness of echoes). One thing this analogy brings out, I think, is that although the present is a beginning, it is not, strictly speaking, what is preceded by nothing: it is preceded by the *there is* (hoot), a past it has without being. It is to that non-serial 'precedence' that Levinas refers when he writes: "in the instant itself . . . something, *so to speak*, precedes the instant" (EE, 131; emphasis added).

"Carpe diem!", said the sage for three good reasons: because we *are* the present, because we are not the past, because we are not the future. This might seem to be an injunction to embrace a void, since "[w]e have no present, it slips through our fingers" (EE, 167). Such is the ancient paradox of the present: it is everything, because we are it, and nothing, because we do not possess it. The style of thinking about time that begins with Bergson or William James tries to resolve this paradox by denying that we *are* the present. The fact that we do not possess the present is no great loss, so the reasoning goes, because it is not in the present that we *are*. We are our whole ex-istence, where past and future are given co-originally with a limiting instant which binds and separates them but has no existence apart from them (EE, 168). As against this, Levinas's philosophy of the instant aims to vindicate an "unsuspected" (EE, 42) sub-ecstatic dimension of subjectivity wherein "human existence involves an element of stability – it consists in being the *subject* of its becoming" (EE, 168). Though the present is unpossessable, "It is nevertheless in the present that we are and that we *can* have a past and a future" (EE, 167; emphasis added). This present that we *are* is not the always-vanished present of serial time; it is the present that begins before time, the present empowered to be a subject of time.

In displacing ex-istence (notably Dasein's) from the core of subjectivity, Levinas is taking issue with the contemporary expression of the tradition of rescuing subjectivity from the system of worldly determinism by denying it substantiality, locating it 'anywhere out of the world', in some transcendental ego or Bergsonian deep self. It is Levinas's contention that the subject thereby saved is so de-substantialized as to be scarcely worth the trouble: modern philosophy has "sacrificed to the spirituality of the subject its very subjectivity, that is to say, its substantiality" (EE, 168). The self is for Levinas substantial in a strong and straightforward sense: "we do not believe that consciousness needs things in order to exist, in the way things need consciousness" (TIPH, 80). But the crucial consideration here is that this substantial self, though it exists before the world, does not exist outside time; it is temporal, contrary to traditional opinion. So Levinas's task is to exhibit, as fundamental, the time of this pre-worldly substantial self, to tell us "how we are to understand subjectivity [as substantial] without, however, locating it outside of becoming" (EE, 169). To this I now turn.

From the present to time proper (time-Other-wise)

The present recommences *ad infinitum* as the event of identification of self (*moi*) with self (*soi*): the self is alone with itself. Hypostasis is a dialectical event involving the sameness of being as thesis, a "break" (TA, 32) from Being as antithesis, a "dialectical return" (TA, 32) of sameness to sameness as *self-identifying* sameness. This dialectical return has nothing to do with retention, which it precedes. The burden of selfhood is not *originally* the burden of the retained past. The self produced by hypostasis is a "reference to self *within* the present" (EE, 150; emphasis added). We suffer in the instant, or rather suffering originally *is* this stagnating in windowless Being-self.

The experience of the burden of selfhood, the unhappy "solitude of the couple" (TI, 151) *moi* and *soi*, implies the thought of liberation from selfhood, but the thought falls short of the reality: "we cannot derive from the experience of servitude the proof of its contrary" (EE, 152); which means that the unhappy couple have no dialectical grounds for divorce. Only an external event supervening disruptively on the inner logic of their situation can answer to their desire to split. That event is the relation with the Other, who places the self "elsewhere" than with itself, cleaves the *moi* from its *soi*. To be saved from Being-self is to be rendered "exterior to the self on the basis of the Other", to be placed by the Other 'away from home'. Levinas's amazing proposition is that this liberation is the temporalization of fundamental time. He is saying much more than that 'other people take us out of ourselves', he is saying that the face of the Other opens up a time-dimension (time proper) of absolutely fresh instants. The event of being taken out of the self by the Other does not happen in some pre-established time – it *is* the 'event' of time. Time's infinity expresses the infinity of the Other's ethical value: time is not, *contra* Heidegger, a dimension whose finitude is grounded in being-for-death (TI, 260).

Passing time – the species of serial time in which the present undergoes modification – does not save the self from itself. All suffering possesses a "definitiveness which its evanescence does not undo" (EE, 159). Husserlian evanescence cannot dissolve the suffering bond of self with self because the now-phase recommences as a past *present*, the future is what the present will be – a "future *present*" (TA, 64). In this "time of the Same" the present refuses to die. The passing of time merely lengthens the chain which binds being-self to itself.

Nor does the world save the self from self: we cannot 'lose ourselves in the world'. "This transcendence through space does not get us out of solitude" (TA, 52; see TA, 73). While it is true on the one hand that

> in everyday existence, in the world, the material [self-adherent] structure of the subject is to some extent overcome [in that] an interval appears between the self and the self [so that] the self does not return to itself immediately
>
> (TA, 54)

it is nevertheless a fact, exploited by idealism, that the *solitary* subject never encounters in the world anything absolutely different from itself. The case is not essentially different on the assumption that our fundamental relation to the world is, as Levinas claims, the non-objectivating intentionality of enjoyment rather than the objectivating intentionality of knowledge. For enjoyment is already a rudimentary idealist "knowledge" in the sense that the enjoying subject draws into itself the nourishing sensory data in which it is absorbed – it is "absorbed in the object *which it absorbs*" (TA, 52; emphasis added). The enjoying self finds *itself again* in its 'food', so that its being-in-the-world involves "not at all a disappearance of self but [just] a forgetting of self" (TA, 52); "it is with myself that I find myself again in knowledge and enjoyment" (TA, 47).

As I indicated, Being-self is for Levinas the essence of suffering, whether as physical or mental torture, or just mild boredom. A useful distinction here, not made by Levinas, would be between pain and suffering. People sometimes inflict on themselves greater pains just in order to *distract* from lesser pains, which suggests that the suffering essence of the pain from which the greater pains distract consists in the self's *adherence to itself* (in pain). Likewise, a pain I can bear as long as I believe I can stop it at will (must not adhere to it) often becomes unbearable when I discover I cannot stop it (must adhere to it). The essence of pain is suffering, which is not something physical or mental, but the incapacity of the self to separate itself from its (sometimes pained) self. Physical and mental pain are contingent specifications of suffering, whose essence is the self's knottedness into itself: "the content of suffering is of a piece with the impossibility of detaching oneself from suffering" (TA, 55).

This impossibility is the unavailability of time – the fact of the self importuning itself in standing inescapably in its in-stant. This self-importuning is what delights the torturer – watching his victim pain himself, as if he (the torturer) were not causally responsible. Suffering is *mine* in virtue of the same twistedness-into-itself ("torsion", EE, 154) that binds the self/instant to itself and makes it individual. The remarks developing this train of thought in the closing pages of EE constitute a meditation on suffering in relation to time unequalled in profundity (and difficulty) by anything in the philosophical literature.

The deepest demand of suffering is not that it become past in "the time of the Same" which merely lengthens the chain binding self to self. It is that suffering "be repaired in its very present" (EE, 154). Whatever else it is, this is not the pointless demand that what is done be magically undone: "The demand does not concern . . . strictly speaking the impossible destruction of that presence" (EE, 159). What is Levinas on about? What in particular has the satisfaction of the deepest demand of suffering got to do with time and with other people?

The deepest demand of suffering is a "hop[ing] for the present" (EE, 156), suffering "hopes for the present itself" (EE, 158). How can we 'hope for the present'? Has Levinas not admitted that the present is indefeasibly the way it is? Is the most we can hope for not that suffering will cease and be compensated?

Recall that for Levinas suffering is essentially the unavailability of time. Since a future instant is therefore a *non*-prospect, suffering can only hope blindly for salvation (which in the event *turns out* to be granted by the advent of an absolutely new instant). But if this new instant detaches from the present, as required, how can it 'repair' it *in its present*? Does the detachment not simply leave the suffering present behind, forever unredeemed, forgotten? And would this ethical outrage not entail, at the consequential metaphysical level, an untenable atomistic conception of time? On the other hand, if the new instant is not a present that detaches from the present, then it is not absolutely new and affords no remedy. We are back with the time of the Same. Can there be a third way between Cartesian atomism and Husserlian continuism? The deepest demand of suffering is at any rate the demand for that third way.

There can be a third way if the new instant emerges as a *resurrection* of the suffering present instant; that is, if the new instant is the suffering instant *again*, only *not as the same again*, inasmuch as absolved from its suffering. This "fresh instant" (EE, 156) is the coming-again, as ab-solved, of the *present* obtaining remedy in its presentness. This paradoxical unity of *again* with *otherwise than the same again*, which defines resurrection, is Levinas's original time: a multiplicity of instants standing to each other in the relation of ethical ab-solution, as distinct from in absolutely no relation. The face-to-face with the Other is the instrument of the self's death (detachment) and resurrection. We touch here on the ethical core of Levinas's metaphysics of time – but I think we would do well to go back over some of this tortuous ground.

Suffering's hope for the present turns out to have been the hope that the present will 'come again', as something "totally surprising" (TA, 64), unforeseeable, as an absolute future which is not the future *of the present* and which does not become a past future – "does not fall back on the past which it was destined to renew [but] remains an absolute future" (TI, 249). The idea of time (time proper) comes to suffering as an answer to the suffering present's hope that its present "will benefit from a return" (EE, 156): "time is the answer to hope for the present" (EE, 157). Hope for the present is answered by the advent of the ab-solving future "undoing of the knot which is tied up in [the present]" (EE, 159). Suffering turns out to have demanded "the absolute alterity of the other instant" (EE, 160). In gifting this, time justifies itself, though we cannot say that the demand of suffering *was* for time: the demand made by suffering has no inkling of time prior to its 'advent'.

The absolute future "adds novelty to being, [it is] what is absolutely new" (TI, 260) because it absolves from self/sameness. This conception of the absolute future is central to Levinas's enormously subtle analyses of *eros* and paternity (respectively TI, 247–251, 255–260), which are ways of transcendence, of living in 'time-Other-wise' (time proper). Limitations of space prohibit my dealing with these analyses here, beyond remarking that they carry out the project, outlined in the closing pages of EE, of describing ways in which my relation to the Other opens up time proper as a "dimension of fresh air" (EE, 156) in which the self can be, without fragmentation, absolutely exterior to itself.

"Fresh air" alludes to how the novelty produced by time-Other-wise differs from the Bergsonian variety. It evokes a heady void, the death of the instant, thanks to which "the deep work of time delivers from this past"; whereas "the novelty of the [Bergsonian] springtimes which burgeon in the heart of the instant, resembling, as in good logic they do, what has gone before, already burdens itself with all those lived springtimes" (TI, 160). The moral renewal that Levinas finds in time requires more in the way of temporal novelty than is delivered by the Bergsonian re-creation of the freshness of the past. The weight of *durée* which Bergson's deep self drags behind itself blights any hope of liberation from self. There is no more fresh air in Bergson's *durée* than in Husserl's 'constipated' time-consciousness. Bergson's moment never dies, his is "a philosophy without death" (TA, 72).

On the other hand, a philosophy of time with too much 'death' (over-exposed to fresh air) would seem to raise the spectre of temporal atomism, which we might be tempted to read into the following bald remarks:

> Resurrection is the principal event of time. There is therefore no continuity in being. Time is discontinuous. One instant does not arise out of the other without interruption, by an ecstasy. The instant in its

> continuation – finds its death and resuscitates. Death and resurrection constitute time.
>
> (TI, 260–261)

But this is not atomism. Viewed in the light of the tradition, the originality of Levinas's position is that it yields what phenomenological philosophy of time since Bergson had been teaching was impossible, namely a conception of temporal discontinuity untainted by atomism. The difference between atomism and Levinas's non-atomistic discontinuity is the difference between the proposition that the present stands in absolutely no relation to the past and the proposition that the present stands to the past in the ethico-temporal relation of ab-solution. The relation of ab-solution consists in the breaking of a relation, in "a rupture of continuity and continuation crossing the rupture" (TI, 160). Atomism and continuism are equally false of fundamental time: "Although time does not bring about a succession of moments, indifferent to each other, of mathematical time, it can no more truly be said to be the *continuous durée* of Bergson" (TI, 260).

But as we have seen, Levinas denies more than that fundamental time is a mathematical succession – he denies that it is any kind of succession. When he speaks of "the absolute alterity of the other instant" he does mean 'other', not 'next'. A return (resurrection) does not generate a succession (nor complete a cycle, since it is not as *the same* that the present returns). On the only occasion when Levinas does allow himself the expression 'the following instant', the indication that it is a makeshift is clear: "[w]hat is called 'the following instant' is the cancellation of the irremissible involvement of existence fixed in the instant, the resurrection of the 'I' " (EE, 157).

This talk of resurrection and absolution is metaphorical in so far as it does not refer to bodily resurrection after death or to a rite administered by a priest in accordance with fixed formulae. Yet insofar as 'resurrection' and 'absolution' denote a liberation from evil and a new beginning that is also, paradoxically, a return, Levinas is using these words literally. He is not using them analogically to describe a conception of time that is not literally ethical. For his conception of time is essentially ethical. Thus when he says that the mutual exteriority of the instants of Time-Other-wise is "absolute" he seriously means that by its "absolute alterity" each instant morally absolves from the evil of Being-self.

Time-Other-wise as a condition of knowledge

The production of separation

"The production of separation is linked to time."

(TI, 144)

The theories of the knowing relation on offer in the tradition record capitulations in the face of the paradoxical tension between the world as existing for me and as existing independently of me. Absolute idealism capitulates by absorbing the world into the self, materialism by doing the opposite. The first offends against our animal instinct that the world is something radically different from me, while the second leaves us wondering how I can know a material world of which I am said to be part. Kant's blend of transcendental idealism and empirical realism satisfies common sense no better, since the realist element is the *unknowable* noumenon. We want knowledge to be the achievement of knowing objects as they are in themselves, independently of subjects. That is our paradigm of knowledge. But is this not demanding the impossible? How can the knowing relation be a relation to what is outside of that relation?

The mentioned paradigm is at the centre of the somewhat submerged theory of knowledge we find in Levinas: "knowledge is a relation with what remains outside all relation" (EE, 148). What makes Levinas think he can have the impossible? Well, the first point to note is that "a relation to what remains outside all relation" also describes the ethical relation to other people. So we have here the suggestion of a homology between the epistemological relation to alterity in general and the ethical relation to the Other.

I understand it to be Levinas's proposition that this homology is symptomatic of the epistemological relation to alterity in general being transcendentally *grounded in* the ethical relation to the Other: no Other, no alterity in general. "The total liberty of the Same [of self] in representation has a positive condition in an Otherness (*Autre*) which is not something represented, but other people (*Autrui*)" (TI, 98). It is consistent with this proposition that a dual reading, epistemological and ethical, is invited by such statements as the one just quoted, or by "[s]eparation opens up between terms which are absolute yet in relation, [terms] which absolve themselves from the relation they maintain" (TI, 195; see also TI, 21, 169). This relation might be read as paradigmatically epistemological, but also as characterizing a conversation between a man and a woman where each takes pleasure in leaving the other's autonomy totally 'untouched', absolute: pleasure in communication as distinct from cloying contact. Thus Levinas, writing the point large:

> the freedom of other people will never have been able to begin with my own freedom; that is to say, be contained in the same present, be – or through reminiscence re-become – contemporaneous with my own freedom, be representable by me. The unlimited responsibility in which I find myself [engaged] comes from hither-side of my freedom, from a non-present *par excellence*, from what is non-originary, from the an-archic, from hither side of, or beyond, essence . . . Responsibility for the Other does not date back to a beginning: my relation to another freedom cannot possibly be contained in a decision of my freedom . . . it is before the origin. An-archy.[9]

Other people are for Levinas 'absolutely before me', in both the 'temporal' sense of absolutely past relative to my present encounter with them and in the ethical sense of 'to be preferred unconditionally to me'. The temporal sense is grounded in the ethical: the Other is absolutely past (unrepresentable) *because or* inasmuch as my responsibility for him/her has no beginning in a choice of my will. Now, objects exist before my knowledge of them in the same temporal sense: as a knower I stand to them in a relation that is non-relativizing, a relation that is the breaking of a relation, a relation in which the object achieves its exteriority or transcendence by absolving itself from the subject; what is to say that the paradigmatically known object is no more present than the Other is.

This contests the way common sense understands time to be involved in ordinary empirical judgments of mind-independent existence, to wit, that a thing judged to exist is co-present with my judgment but precedes and outlasts my judgment in serial time. As against this, Levinas implies that the object of judgment is *absolutely past* relative to the judgment, not co-present with it. In *that* sense "every distinction between perception and the perceived [the production of exteriority] rests on time, on the non-synchronicity between the aiming intention [*la visée*] and what is aimed at [*le visé*]"; which means that the eye of intentionality opens on an object which is "older than the intention".[10]

This claim that time, as the above-mentioned non-synchronicity, is the medium of epistemological separation between subject and object naturally evokes the counter-claim that the medium is space. Are knowable objects not separate from me in virtue of being 'out there'? Not according to Levinas:

> the void of space is not the absolute interval out of which an absolutely exterior being can surge up ... Vision is not a transcendence ... It opens nothing which, beyond the Same, would be absolutely other, that is to say, in itself.
>
> (TI, 165)

We noted him hint at this some way back, with his denial that being-in-the world unsticks the self from itself.

Let us remind ourselves that Levinas's complementary positive thesis is that "the relation with other people alone introduces a dimension of transcendence" (TI, 167); or that "the light of the face is necessary to separation" (TI, 125). In arguing this he agrees with Heidegger that the relation of the eye to a thing stands within the deeper relation of the thing to a general openness that is not a thing and that conditions the appearing of all things. A known thing "stands within an opening on a being which is not a *being* – which is not a 'something' – but what is necessary in order that, generally, a something should manifest itself" (TI, 164).

What is this opening or light without which no thing can show up to a seer as absolutely exterior? Space might seem to fit the bill rather well – is space not grounded in Being's light, and so a "chasing away of darkness" (TI, 163)? This Levinas denies, on the ground of his own conception of Being (the *there is*) as darkness (evil). The claim of vision to encounter an absolutely external thing has always rested, he thinks, on the idea that "a being comes from nothingness": "this coming from a void explains its privilege of objectivity and its presumption to coincide with the very being of entities" (TI, 164). The idea is that if things came from anything other than a wholly diaphanous void – if they came from *something* non-appearing behind them, such as a noumenon – then they would not be grasped 'in their very being'. But it must be false that the diaphanous all-giving void is space if it is true that space is grounded in a Being which is dark. For then space will not be a wholly diaphanous space-light, the appearing of things *as* spatial will merely conjugate the dark verbality of the *there is.* According to Levinas, space-light is in fact troubled in its inmost essence by the dark fact that *there is* space-light. Space-light is porous to the darkness it banishes:[11] "the light", in banishing the darkness, "does not stop the ceaseless play of the *there is*. The void produced by light remains an indeterminate thickness" (TI, 165). Space-light is thick with the *there is*, which means that it "is certainly not equivalent to an absolute nothingness" (TI, 165) or diaphanous void. The darkness here is not of the provisional kind that hangs around in the unseen sides of seen objects, in the blind-spots of theories, and which can be dispelled by looking at the object from the other angles or by further theoretical research. The suggestion is rather that light harbours its own non-relative opacity "of another order" (TI, 165) from the provisional opacity.

Levinas's thesis of the *definitive* opacity of space may be put in terms of the way vision sees across space towards a horizon from which it beckons the hand to movement and contact: "vision transforms itself into grasping" (TI, 165). Visual consciousness stretches over a finite [crossable] distance to a horizon, in returning from which [to the hand] it locks space into in the "play of the Same"; this is because the movement towards the horizons is not "a breaching of the horizon" into what is absolutely other; it does not enable us to encounter things "from a standpoint beyond all beings" (TI, 166; see TA, 47, 53); it does not place the subject's relation to the object within the subject's relation to absolute alterity: "the [spatial] light does not allow us to broach the face of things" (TI, 165–166). How, then, is vision possible, given that we cannot take up a point of view in the diaphanous milieu beyond being? Levinas appeals to the heroic forgetfulness involved in the resolve 'to enjoy life all the same':

> the possibility of vision in the light is precisely the possibility of forgetting [the *there is*], of approaching objects *as if* from their origin, from nothingness. This exiting from the horror of the *there is* is announced in contentment and enjoyment.
>
> (TI, 165; emphasis added)

Where does this leave Kant's claim that space is a form of separation? Through their common relation to the hand, objects are tied together into spatial relations, of which the Kantian intuition of space is the apriori form, according to Levinas: "empty space is the condition of this relation" (TI, 165). Space is a form of separation, as Kant claimed, but it separates objects *from each other*, not from the subject. "Space, instead of transporting [the self] to a beyond, simply supplies the condition of the *lateral* meaning of things [grasped] within the Same" (TI, 166). So the void of space does not supply the sheer light requisite for the absolute separation of self from non-self, as required by the paradigm of knowledge. Only diachronic time-Other-wise supplies that light.

The spatial light remains troubled when it is the 'second' light of geometrical intuition, as distinct from the 'first' light of sensory perception. Though a geometrical ideality is 'light' relative to the physical shapes which exemplify it, it is also itself "a plenitude", something "in turn *seen*" (TI, 164); which means that "we need a light to see the [geometrical] light" (TI, 166). The concepts of intuitive (Euclidean) geometry involve a certain 'thickness' because they are limits of idealizations from *seen* (therefore existent) things (as Merleau-Ponty had also claimed):

> the line is the limit of a thing, the plane is the surface of an object . . . the notions of geometry are compelling on the basis of 'a something' . . . light-space involves the diminution to the point of nothingness of these limits.
>
> (TI, 164)

Geometrical concepts are admittedly prior in meaning to the objects seen as exemplifying them, but they are prior only *after*, and as conditioned by, the sensory seeing of surfaces. The intellectual seeing arrogates priority to itself by 'topping' the sensory seeing which conditions it. The intellectual seeing is "anterior after the event" (*antérieur postérieurement*: TI, 144). As we shall continue to see, this usurpation of anteriority – "the posteriority of anteriority" (TI, 25) – characterizes, for Levinas, the general temporal structure of mind in all its sub-ethical (egocentric/intentional) operations.

The temporal structure of being-at-home

This posteriority of anteriority is epitomized by the ambiguous status of the home. For is the home not a power-base *vis-à-vis* the natural environment *on which it stands and depends*? The home is extra-territorial, a domain of sovereignty carved out in an alien environment. Though Levinas has analysed the concept of being-at-home (dwelling) in a literal sense, I believe that it works as a pervasive, almost necessary, metaphor for his understanding of the time-structure (posteriority of anteriority) of the sub-ethical self. And with this I commit myself to the exegetical thesis that this sub-ethical time-structure (time before time-Other-wise) is inextricably bound up with notions of power – as I hinted in calling his present a power-base. For it is only once anteriority has been identified with 'power-over' and posteriority with dependency that it might make sense to claim that the home vs environment difference corresponds to (a convolution of) the before–after difference. On any ordinary view of the before–after relation, the home is neither before nor after the environment.

We saw earlier that Levinas thinks of hypostasis as the event through which the game of being a subject begins. It will be worth going back briefly to this event to confirm that the game of being a subject is actually a power-game.

At-home-in-the-self: posterior anteriority

As I have said (p. 79), the adventure or "intrigue" thematized by Levinas is pre-intentional. This means that "to assume existence is not to enter into the world", or that "the inscription [of the self] into being [as being-self] is not an inscription into the world" (EE, 173). It is an 'inscription' into the substantiality of thought, this being construed by Levinas in a way that is at variance with two stock philosophical positions on the relation of thought to cognitive substance: (i) that thought is, as for Descartes, an attribute grounded in a logically prior self-substance; and (ii) that thought is devoid of substantiality just because it depends on no Cartesian substance and must 'borrow its being' from its objects, as Sartre claims. Levinas rejects both these positions. In hypostasis, thought positions itself *behind* its determinations through an act of self-identification in which it denies of itself that it is its determinations. The substantiality of thought is not a sort of something, but rather this power of contentless self-identification: thought "has identity as its content" (TI, 5), is a "being whose existing consists in identifying with itself" (TI, 5). The substantiality of thought is the accomplishment of thought itself; it is thought's "power of infinite withdrawal, the power to find itself again behind what happens to us" (EE, 87). This kicking-out of thought to behind its episodes "accomplishes the condition of all interiority" – it is an event "behind the cogito" (EE, 147) in that it is presupposed to thought as a centrifugal transaction with the world (EE, 139, 171). The first movement of thought is therefore *away* from what-is/happens to a position, or rather 'positionedness', from which it returns (as intentionality) to the world, which means that "transcendence is not the first initiative of ontology" (EE, 172). This fits well with our intuition that, as Levinas says wisely, "consciousness never has its back up against the wall" (EE, 116) – but rather against an "exit door" (EE, 171). Thought goes through that door into its 'home' in the present, which is puffed with power.

In the name of what does thought kick out? Not, ultimately, in the name of its autonomous self. For we have seen that in identifying itself with itself thought only ensnares itself in the intolerable finitude of being-self, from which it desires to kick out in turn. I think we catch a glimpse here of "the idea of the infinite" (the Other) magnetizing the ontological adventure from a pole outside the world. So the fact that the Other is not encountered in the world means that the world does not bear witness to her: the world is haunted by the Other as by the light from the sun beneath the horizon.

At-home unhoused in the world: posterior anteriority strengthened

We could accomplish hypostasis while hurtling through interstellar space with no sense of our bearings. What we would feel the lack of in this circumstance is the "amplification" (EE, 80) or "confirmation" (TI, 128) afforded by *terra firma* of the power-base or posterior anteriority achieved in hypostasis. To be-in-the-world is, to be-at-home, again, only as more stably positioned than in hypostasis. It is to have dug in one's heels, not to have 'fallen' in Heidegger's sense.

Our fundamental way of being-at-home-in-the-world is much the same as a puppy's. What happens when we open our eyes? Only a mind corrupted by philosophy would answer: 'we become aware of objects displaying certain phenomenal properties and standing in spatio-temporal and causal relations'. Levinas wants to remind us of the half-forgotten truth that what immediately happens when we open our eyes is that we find ourselves enjoying the spectacle (TI, 103). Imagine yourself eyeball to eyeball with a black cat sporting long whiskers. Nothing outrageously beautiful, but 'something to look at'. Is this situation not naturally expressed by saying that your eyes 'rest' on the cat, that seeing is an intentional attitude supported by what is seen, "nourished" by what "from another point of view may appear as an object of thought, as simply constituted" (TI, 109)?

To be-in-the-world is *fundamentally* to live from its elemental qualities, and that is to be at home in the world: "living from the world [the self] lives at home" (TI, 121); "the happiness of enjoyment affirms the Me at home" (TI, 117). Living from the world is a "primary pleasure [which] does not alienate the self, but sustains it, constitutes its home" (TI, 116–117). There is an "intentionality of enjoyment", a relation to the world as a sensory plenitude prior to and independent of the theoretic/objectivating intentionality privileged by Husserl to the bitter end (as Levinas argued in TIPH). We do not first represent things to ourselves in a theoretic and affectively neutral way, then proceed to enjoy them in an intentional act adding a founded layer of affective meaning. The world is enjoyed before it is represented.

Enjoyment is even prior to the instrumental intentionality which Heidegger and his followers took to be basic. "Everything that is given in the world is not a tool" (EE, 65). In the first place, it is just false that a house, for example, is 'an instrument for dwelling in' and that a soldier's bed, his food and clothing, are *for him* (as distinct from the Ministry of Defence) so much 'equipment'. They are ends in themselves, refuges from instrumentality. In the second place, even when things *are* used as instruments, our enjoyment of their materiality encompasses our handling of them. Enjoyment seeps up into the instrumental attitude: we enjoy the feel of a well-balanced hammer, the resistance of good garden fork, even as we use these things. We do not normally use things *instead* of enjoying them (unless in conditions of oppression: TI, 84). While it is true that eating and working are means of sustaining life, it is also true that they naturally blossom into ends which we pursue for their own sakes: we do not just eat and work to live, we enjoy eating and working to live. Instrumental action spawns autonomous value 'hither side of' its goal: "the means is immediately sought as an end and the pursuit of this end becomes an end in itself. Thus things are always more than bare necessity, they constitute the charm of life" (TI, 84). Doing something to some purpose is not an unswerving movement to just one appointed goal. The sensory contacts involved in striving to sustain life are absorbed into that striving as value-elements accruing to it, so that we are always "beyond the being in which things are carved" (TI, 85). Life is not lived 'on the bone', it is a 'second-degree' affair in which being is permeated with value acquired on the way.

Enjoyment exists both in the pure sensory state, apart from action and thought, but also as the general atmosphere of thought and action, which are founded modes of enjoyment. Just as we enjoy using tools, so we enjoy thinking: "[b]ehind theory and practice is the enjoyment of theory and practice" (TI, 85). Human life is basically aesthetic, not theoretic or pragmatic. The most efficient practical action is stylish, the best objectivating representations (scientific theories) elegant (EE, 86–87).

In anyone over-exposed to existentialist doom and gloom, Levinas's descriptions of the intentionality of enjoyment will produce the effect of a breath of fresh air devastating with hurricane force the bleak landscapes of Heideggerian *Angst*, wreaking havoc with the "interpret[ation] of the world as a horizon from which things present themselves as tools, as the equipment of an existence concerned about its being" (TI, 137). Concern with Being or with selfhood, with equipment – the whole edifice of notions congruent with the primacy of horizonal time – collapses under his gentle reminders that in our first encounter with the world we are not in the least concerned with our own being, nor even with physical self-preservation (see above pp. 77–78). Our primary enjoying relation with the world is terribly sincere (EE, 56; see EE, 67), innocent of ontological ulterior motives. Eating contents itself with satisfying hunger, aspires to no further end. "Man loves his needs" (TI, 120) because he enjoys satisfying them.

What matters in all this from our point of view is the conjunction of the claim that enjoyment is the basic mode of being-in-the-world with the claim that its time is quite different from the horizonal time of Husserl, and the horizonal-ecstatic time of Heidegger. The fact that the time of enjoyment is "anterior to the time of representation" (TI, 114–115) means that the temporally-grounded opposition between the finite and the infinite (between now and its open time-horizons), which obsesses Husserl's representational consciousness, is unknown to enjoying consciousness; in our first encounter with the world we do not understand an object-in-itself (as distinct from how it appears to me) to be a presumptive totality located at the ideal limit (horizon) of its in-principle infinite series of mutually confirming appearances. "Objects *satisfy* me in their finitude without appearing against a background of infinity" (TI, 108; emphasis original). So at this hedonistic level we have no split between an object's appearing and its being. The objects of the intentionality of enjoyment[12] differ from those of representational thought (Husserl's 'perception') in not being appearances *of* something which infinitely surpasses them: they are "appearances without there being anything that appears" (TI, 109). The intentionality of enjoyment offers "freedom from all the implications, from all the prolongations [horizons] offered by thought" (TI, 112).

Strictly speaking, the intentionality of enjoyment does not grasp *objects* at all; it is not a transaction with discrete substances. Immediate perception "bathes in the elemental" (TI, 104) – earth, air, water, etc. – wherein no substances are discriminated. Our senses, on encountering the world, are flooded by sheer free-floating formless quality, detached from supporting substances: "sensibility places us in a relation with pure quality without support, with the elemental" (TI, 109). Qualities present themselves as "adjectives without substantives" (TI, 105), as "content without form" (TI, 104). The elemental environment is thick, dense, unfathomable, but this is a different sort of depth from that structured by the split between being and appearing. The elemental is absolute appearance, it "comes from nowhere" (TI, 114), is without beginning or end. By contrast, the depth of a thing represented as a substance resides in its being surpassing its appearing. And the appearances of a substance 'come from' the substance itself: the visible and invisible sides of a vase are equivalent inasmuch as possessing a common ground. The elemental resists exploration by our sensory faculties, which is why it is ominous; science can find out more about the wind, sky or sea, but the perceiver cannot enhance his or her perceptual purchase on them by 'looking at them from the other side'. For they have no other side.

The enjoying self is "famished" (TI, 102) and "lives from" its sensory world – 'eats' it. Taking eating as synecdochic for enjoyment as a whole "is justified by the place it occupies in everyday life, but especially by the relation between desire and satisfaction which it represents and which constitutes the very essence of life in the world" (EE, 65). However, "[t]hat from which we live and which we enjoy is not fused with life itself" (TI, 94); the intentionality of enjoyment constitutes the world as exterior. But now we seem to have this problem: taking eating as a key synecdochic figure for being-in-the-world seems to be at variance with the world's alleged exteriority, in that a certain usage of the alimentary trope (which Levinas traces back to Plato: TI, 86) has done long service in the very philosophical tradition which reduces alterity to the sameness of self,[13] the acquisition of truth to the satisfaction of hunger, knowledge to presence. The question is: how is

Levinas using the alimentary figure such that it squares with the exteriority of the enjoyed world, with his statement that "we oppose throughout this book the total analogy between truth and food" (TI, 86)? If enjoyment is a rudimentary knowledge, as he claims, is this knowledge not a form of 'digestive idealism'? If it is, the knowing subject would be *absolutely* anterior to the world, not anterior to it 'posteriorily'. So we might seem to have encountered here a difficulty for my claim that with Levinas posterior anteriority is the temporal form of all sub-ethical operations of mind. But I think not.

Notice that Levinas says that the analogy between food and truth is not total, not that it is totally inapplicable. Taking our cue from this, we might surmise that the difference between his own and the idealist application of the alimentary trope might be this, that although to know is to eat, it is not to eat *up*, which is what idealist knowledge figuratively does to its object. But now we have the suggestion that the question, What then *resists* the 'famished bite' (constitution)? is in order. And this would put us on the wrong track, for the fact that sensibility "satisfies itself with the given, contents itself" (TI, 109), means that *nothing* resists the famished bite. Levinas is impatient with the reason offered by some phenomenological existentialists for limiting the scope of constitution, namely that it is checked by an irrational bedrock of facticity which cannot be converted into a sense or noema:

> if the intentionality of 'living-from' is not constituting it is not, then, because some ungraspable inconceivable content, which would supposedly be unconvertible into the sense of a thought and consequently unrepresentable, would compromise the universal scope of representation and of the transcendental method.
>
> (TI, 102; see also TI, 120)

His positive point is that in "living from . . ." the movement towards the object is not checked by residual facticity, but *reversed.* Our business now is to see how this reversal conforms to the convoluted temporal structure of non-paradigmatic knowledge (sub-ethical mind): the posteriority of anteriority.

In the phenomenological literature, the subject-body is described as an organizing centre of perception, especially by Merleau-Ponty and Sartre. Levinas, however, wants to convince us that the subject-body is a secondary centre whose centrifugal movement towards the world is subtended by a countervailing conditioning movement from an absolutely older centre, which is the elemental world; in going out intentionally towards the world, the body constitutes the world as exterior, but this going-out is a *return* to an absolutely past anonymous centre which "nourishes" the going-out movement itself, which means that the going-out movement "comes from the point to which it goes" (TI, 102), or that in going out it makes itself "interior to the exteriority which it constitutes" (TI, 102): "in nourishment the self rejoins itself" (TI, 120). The centrifugal movement is not checked by facticity, but rather "the [power-] game changes direction" (TI, 102). Enjoying intentionality is conditioned, in its activity of conditioning the world as exterior, by an energizing centre older than enjoying intentionality and to which that intentionality 'goes back'. What the subject contains as represented is also what supports and nourishes the subject in its constituting activity; which means that that to which it goes back precedes the constituted (exteriorized) world, which has always-already "reverse[d] itself [*vire*] into a past which the present of representing has not lived through, as an absolute past which does not receive its sense from memory" (TI, 103). The original centre is therefore an anonymous "non-representable antiquity" (TI, 111), "[t]hat which is represented, the present, is *made* [*fait*], already past" (TI, 103).

These descriptions are so many ways of saying that the anteriority of the enjoying self is posterior. This posterior anteriority is midway between absolute anteriority and absolute posteriority, between the limit of outright idealist 'power-knowledge', on the one side, and the submission to the 'thing itself' characteristic of paradigmatic realist knowledge, on the other. In speaking of "the *gentle* mastery of enjoyment" (TI, 114; emphasis added), Levinas alludes to that intermediary degree of the subject's mastery (or, correspondingly, of exteriority); because the subject's going-out to the world to constitute it as exterior is at a deep level, its going-back to rejoin itself in its immemorial pact with the world, the distance separating it from the world is traced across a common ground, whence it follows that "between the self and that from which it lives the absolute distance separating the Same from the Other does not extend" (TI, 116). Or put differently, being-in-the-world in only a midway stage in the ontological adventure.

That enjoyment is an accomplishing of the present instant against the background of the absolute past (the *there is*) is implicit in Levinas's claim that being-in-the world is an amplification of hypostasis. And indeed Levinas says often enough that there is an instant of enjoyment (T1, 117, 119, 124). Yet this does not mean that enjoyment is immured in the present, oblivious of the future. On the other hand, we have seen that the future of enjoyment is not the future of passing time; not an ecstatic future *present*. So what sort of future belongs to enjoyment?

The enjoying self is "sovereign" and forgets the *there is* (TI, 166) to the extent that it loses itself in the nourishing plenitude of its vital world. In that sense it is at home in the world. But as we also saw, a home is essentially extraterritorial (TI, 104, 124), a domain of privacy and security established in an ever-threatening, unfathomable environment: homes are liable to subsidence. In other and more literal words, the vital world is continuous with the *there is*, "prolongs itself into the *there is*" (TI, 116), "loses itself in the *there is*" (TI, 121, 123). This gives rise to insecurity and "worry about the morrow" (TI, 114), which worry, Levinas claims, is "the original phenomenon of the future" (TI, 124). Unlike the Husserlian protentional future, the future of enjoyment does not "go beyond present time in the project which anticipates the future" (TI, 140), "we are not dealing with a representation of the future, but with a time anterior to representation" (TI, 114–115). This pre-representational future lurks *in the present* as a "nocturnal dimension" (TI, 116) or "nothingness" (TI, 115, 116, 120, 124) threatening to cut off the enjoying self's sources of satisfaction. But is this not to say that the future *is* the present – and is that not nonsense?

I think not. A chill wind suddenly pierces the serene warmth of an autumnal evening, and we say to ourselves, 'Winter is coming'. Unmistakably, on such occasions, the chill wind unseasonably intrudes *into* the present *from* the future. There is clearly no question here of a present perception protending *towards* a future perception; for the protentional future is not yet present, whereas the future we have just felt *is* present: we *feel* the invasive presence of the winter we say is coming, we feel the future in the present: "the future is *already in* that pure quality from which the category of substance is lacking (TI, 115; emphasis added). Not in it in the well-blended Bergsonian fashion, but rather as "a heteronomy which occurs within interiority itself" (TI, 126), as a sort of jolt. In that it is a dark anonymous jolt rather than a protention opening up a temporal extension ('light from the subject'), the future of enjoyment involves no time-interval postponing danger – "we are not dealing with a representation of the future where the threat offers a period of grace and liberation" (TI, 114). Being in the present as jolting it, the future of enjoyment is as discontinuous with its present as its present is with its past. This future is a disturbance within the instant, a threat to its power.

The above-mentioned distinction between the future as involving and not involving a time-interval has to be borne in mind if we are not to be flummoxed by such apparently contradictory statements as "paradisiacal enjoyment [is] . . . without time" (TI, 137) and "enjoyment opens up the very dimension of time" (TI, 139). What pure enjoyment is 'without' is the conception of the future as implying an *interval*; this consistently with it opening up a future as a disturbance within its instant. But how are we to understand the shift from the first sort of future to the second? We are told that the first "veers" into the second: "the nothingness of the future veers into a time-interval" (TI, 120); but also that this time-'interval' is experienced as a postponement of danger, a period of grace, not as an abstract temporal extent. Not only is 'not-yet' not the original future, but when not-yet does make its first appearance, it is as 'not-yet-too-late'.

At home in a house: posterior anteriority further strengthened

This prudential future into which we have veered, again discontinuously, is the fact that "the present is not yet". But what does this mean? What can it mean to say that "to be conscious is to be in relation with *what is*, but as if the present of *what is* were not yet fully accomplished" (TI, 140)? Levinas is prising apart those supposedly inseparable twins – the present and what is (going on) – by defining the present in terms of the *non-definitiveness*, for the self, of what is indefeasibly going on now. The prudential present is the non-coerciveness of what-is, its incapacity to totalize the self: "the dimension of psychism opens itself up under the pressure of the resistance to totality which a being [self] puts up against its totalization, it is the event of radical separation" (TI, 24). Non-definitiveness, postponement, is not to be confused with defeasibility: "through time, in fact, being *is not yet*; which does not confuse it with nothingness, but maintains it at a distance from itself" (TI, 137). We have been here before, apropos of the non-definitiveness of the pre-worldly suffering of being-self/present not entailing its defeasibility.

This resistance against totalization is powered by the idea of the infinite (of the Other) which, as Levinas says in a structuring statement, "somehow magnetizes the very field where this production of the infinite is played out" (TI, xiv). Which means: the progressive radicalization of separation maps out the "field" across which the ontological adventure moves towards the production of the infinite as a perspective from which the subject stands in the non-relativizing time-relation to the world requisite for the paradigm of knowledge. But there is no teleology here, the adventure does not progress towards a pre-established goal. A less radical form of separation simply *turns out* to have craved the ontologically subsequent, more radical, form. The self passes though the magnetized field in a series of 'jolts'. The contrast here is with a situation where the self would understand its solitary state in dialectical opposition to a relation with the Other, so that the existence of the Other would be implied in the definition of self, independently of any *encounter* with the Other. But this is what Levinas means to deny with his claim that the advent of the Other and of time-Other-wise involves a jolt to self, "a new event" (TI, 145), irreducible to an implication of selfhood. The jolt is the basis of his resistance to the idea that the Other is involved in the definition of self, whence it would follow that self and Other would be totalized in a system fatal to the Other's infinity or transcendence: "the force of opposition and dialectical attraction would destroy transcendence by integrating it into a synthesis" (TI, 125; see also TI, 24), wherein the ethical relation would be symmetrical (as it indeed is at the abstract level of social morality, where everybody is an Other).

But to return to the veering of the future as insecurity into the future as postponement: "in order that this future can surge up with its meaning of postponement and time-limit . . . the separated being must be able to gather itself together [*se recueillir*] and have representations" (TI, 124). And that requirement "is concretized as habitation of a dwelling or House" (TI, 120). The reasoning seems to be as follows: postponement depends on the fabrication of tools with which to enhance control over the elemental; the fabrication of tools depends on prior conceptualizations of the elemental environment; these conceptualizations, however, cannot be achieved while wallowing in the midst of the elemental – they presuppose the *distance* from the elemental produced by retreating into the dwelling, which affords a further amplification of 'positionedness'. The security of dwelling is the necessary and sufficient condition for representing the world as a controllable domain of reasonably stable substances, as distinct from a welter of groundless appearances. Being the seat of representational thought, of knowledge, the home is itself unknowable – it turns into house the moment we try to represent it.

With his claim that the fabrication of tools rests on prior conceptualizations, Levinas is rejecting Heidegger's doctrine that instrumental or pragmatic intentionality is prior to and independent of theoretical intentionality: "against a modern philosophical opinion which proclaims the autonomy of practice [I claim that] knowledge is the condition of all free action" (TI, 78). Far from representation retracing the "assignment-relations" (Heidegger) or "hodological pathways" (Sartre) furrowed into the world by pragmatic intentionality, the latter gropes its way along a path already beaconed by representation. "The act of representation, strictly speaking, discovers nothing in front of itself" (TI, 97) by which it might be guided. Enjoying intentionality is for Levinas prior to theoretic intentionality, but the latter is prior to pragmatic intentionality. What the defenders of the primacy of pragmatic intentionality forget is that the *fashioning* of tools involves prior conceptualizations of the objects to which they are adapted.

The representing thought (representation) enabled by dwelling understands itself to be "absolutely master of the world" (TI, 121), therefore *absolutely* anterior to it (cf. the gentle mastery and correlative *posterior* anteriority of the nomadic self). But if representation is absolutely anterior/mastering, nothing escapes its grasp; and if nothing escapes its grasp it reduces the object of knowledge to the object of thought (gobbles up the former) and so fails to fit the paradigm of realist knowledge, according to which objects are known as absolutely exterior to the subject. But we know that Levinas thinks we can have non-idealist knowledge of absolute exteriority, and that this is achieved by representation. So we have reason to look more closely at representation's claim to be absolutely anterior/mastering in the dwelling.

According to Levinas, "the intentional relation of representation is distinguished from every other relation ... in this, that in it Alterity does not determine Same (self), that it is always the same that determines Alterity" (TI, 97). But it must be false that representation is universally determining if it is true that it is an intentional relation to transcendence *and* that "all transcendence rests on non-synchronicity". For that means that the object to be known as transcendent must be 'older than', thus escapes the grasp of, the act of representing thought that knows it.

At first sight the notion of representation being universally determining and never determined seems not to deserve a hearing. When I represent to myself the sum of the angles of a triangle, I cannot seriously convince myself that I am contemplating the properties of a black cat. My thought is constrained, surely, by something external to itself. The representation in question on this occasion is "precisely the one that thinks that sum and not the one that thinks atomic weight" (TI, 97). However, just as Descartes' good will gracefully assents to the *compelling* evidence of clear and distinct ideas *without prejudicing its freedom*, so in Levinas's representation

> The object which presents itself to the thinker determines the thinker. But it determines the thinker without touching the thinker, without weighing down on him; in such a manner that the thinker submits to the imposed content of thought 'in good grace', as if he had anticipated the object right down to the surprises it holds in store for knowledge.
> (TI, 97)

This is the representational special case of the already-noted general capacity of thought ("psychicism") to kick out from under, or "hoist itself" (TI, 99) above, all its determinations, thereby escaping totalization. Such is the impudent "geniality" (TI, 99) of representation – what we might call its 'topping reflex'.

But is representation not constrained by its own past, if not by its present content? Is a truth represented by a reasoner at a later stage in working through a proof not determined by truths represented at earlier stages? Not if representation is the feat of re*present*ing, as part of present content, insights we would want to call past if we believed that representation *re*presents, that is, passively acknowledges its incapacity to fetch back its object *as present.* If representation re*presents*, the content of the past is made part of present content and its pastness is forgotten: representation kicks out from what it represents by the same token as it kicks out from simply present elements.

Because it is thus masterful "[r]epresentation involves no passivity" (TI, 98). Where there is no passivity, there is no past, no temporal passage. That there is an indissoluble nexus between pastness and passivity surely belongs to our common-sense understanding of temporal passage: 'the present passes' means pastness *befalls* the present, that becoming-non-present is none of my doing. Whence it would follow that in its own eyes, as masterfully all-determining, "representation is . . . the positing of a pure present without even a tangential attachment to time" (TI, 98).

Now Levinas claims that Husserl's time-consciousness is representational in this sense, which means that it shrinks time to the bare instant and fails to acknowledge the passivity of temporal passage. We might wonder how Levinas can plausibly maintain this in the face of Husserl's insistence on protention and retention. For are these not precisely our 'attachments' to time beyond the instant? And are they not described in a language of passive synthesis testifying to passivity? It is true that Husserl describes retentional intentionality in the vocabulary of passivity, as a "sinking back" or "falling back" into the past. But he also describes it in the active idiom of a "hold[ing] on to" elapsed phases. Levinas interprets this active holding-on-to as a fetching-back which nullifies the passive sinking or falling back, so that the latter is really just the *threat* of passivity, no sooner posed than removed. The tension between passivity (a movement towards absolute exteriority) and activity (a reverse synchronizing movement towards re-interiorization) is resolved to the glory of activity:

> every loss of time, every lapse, holds on to itself, or is recuperated in memory, finds itself again, or is reconstructed; it adheres to an *ensemble* through the agency of memory or of historiography. Consciousness in reminiscence glorifies the ultimate resilience of presence. *The time of consciousness which lends itself to [self-] representation – this is more strongly synchrony than diachrony.*[14]

Whence time as the play of the Same. Or again:

> To represent is not simply to render present 'again', it is to fetch back to the present itself an actual perception which is flowing away . . . to fetch back to the instantaneity of thought everything which seems independent of it.
>
> (TI, 100)

> The Same, in its relation to Alterity, refuses what is exterior to its own instant, to its identity in order to rediscover in that free-floating instant . . ., as conferred sense, as a noema, everything which had been rejected.
>
> (TI, 98)

And since alterity is grounded, as we have seen, in the *non*-synchronicity or "diachrony" of the object *vis-à-vis* the subject, it follows that "[representation is] a mastery, exercised by thinking, over what is thought, whereby the object's resistance as an exterior being vanishes into the object of thought" (TI, 96): "in the intelligibility of representation the distinction between me and the object, between the interior and the exterior, effaces itself" (TI, 96). So if representation is knowledge at all, it is non-paradigmatic idealist knowledge.

To return in the light of this to the situation of the reasoner: he or she does reckon with antecedently intuited truths, but only in the sense that, without being "marked by" (determined by) the past, he or she "*uses* the past as an "objective and represented element" (TI, 98; emphasis original). That is, the representer/reasoner does not understand the truth of present thought to be *subject to* the continuing truth of past thought; but rather the truth-content of past thought is packed into the *present* noema of genially-free representing thought.

After all this, any but the closest reader of Levinas's text would be bamboozled to find him announcing that "representation is conditioned" (TI, 143). How can it be conditioned if it determines alterity without being determined by alterity? Actually, Levinas has issued three warnings that his considerations on the geniality of representation are "true only of the self of representation detached from the conditions in which it latently arises" (TI, 99): "representation is *linked* to a completely different 'intentionality' [enjoyment]" (TI, 98); and "Taken by itself, in a sense uprooted – representation seems to be oriented in a direction opposite to that of enjoyment" (TI, 95). *Seems* to be, but in truth isn't, as we should expect from the earlier recorded claim (p. 91) that enjoyment seeps up into theoretical intentionality; for this means that the relatively passive time of enjoyment must be tacitly unfolding simultaneously with the synchronizing 'powerful' time of representation. Binding these two temporal strands together in two excruciatingly precise sentences, Levinas says:

> Admittedly, the self which conducts its thoughts *becomes* (or more exactly, ages) in the time in which are spread out its [the self's] successive thoughts throughout which the self thinks in the present. But this becoming does not appear on the level of representation.
>
> (TI, 98)

In short, the definition of representation as unilaterally conditioning turns out to have recorded only representation's self-image or "transcendental *pretention*, constantly controverted by life which is always-already implanted in the being which representation claims to constitute" (TI, 143; emphasis added). If representation's claim to be a unilateral and unconditioned conditioning were true, it would be a free-floating "sovereignty in the void" (TI, 143), "an absence of relation" (TI, 143); it would not be a mode of being-at-home-in-the-world. It would die of starvation. But the truth of the matter is that representation is parasitic on the intentionality of enjoyment, whose 'posteriorly anterior' time-structure it in truth repeats (TI, 143–144). Representation therefore conforms to the invariant time-structure of non-paradigmatic (non-radical) epistemological separation. Representation still lives from what it 'knows'.

How can representation discount this dependency, believe in its own presumption, kid itself on? Again Levinas appeals to forgetting: just as the enjoying self forgets the threat of the *there is*, so the representing self forgets its dependency on the enjoying self: "An illusion? – representation is the power of that illusion and these forgettings" (TI, 98). "Representation, interpret[s] itself as eternity" (TI, 98), and timeless sovereignty suffices to make "idealism an eternal temptation" (TI, 144). But in truth it can only "pretend to substitute itself for that life in reality" (TI, 143). The extreme case of representation's illusory presumption to be the measure of all things is its attempt to posit the home as an object of knowledge, to top the vital condition of all representation: "the event of dwelling is without common measure with knowledge" (TI, 126), just because "every consideration of objects, including buildings, takes place on the basis of the dwelling" (TI, 126).

Away-from-home-at-home: posterior anteriority jolted into absolute posteriority

Having claimed that the dwelling is the seat of the representation of the world as objects, and therefore the power-base for the making of tools, Levinas proceeds to the astonishing proposition that dwelling presupposes the presence of the Other in the home. "The passage from instantaneous enjoyment to the fabrication of things is grounded in habitation, *which presupposes the welcoming Other* (TI, 120; emphasis added). "Postponement [in the dwelling] presupposes the absolutely Other" (TI, 5). "The possibility of representing oneself . . . exploit[s] . . . the relation with the absolutely Other" (TI, 95). "Gathering-together presupposes a welcome" (TI, 128). No home without a co-dwelling Other.

But why? Well, given that life is "living from . . .", the question arises why the self, in withdrawing from its immersion in the munificent elements into the home, does not simply cut off its source of nourishment and atrophy. The dwelling self would be a fish out of water if its 'food' were not provided by the welcoming presence of the Other in the home. The presence of the Other in the home is the answer to Levinas's questions: "From where does that transcendental energy come to me, that postponing which is time itself? (TI, 145).

> How can a total reflection [a total dis-engagement from the elemental] be allowed to a being which never becomes the bare fact of existing and whose existing is . . . a living from something? How, within a life which is living from . . . and which is concerned to overcome the insecurity of enjoyment, can a distance be produced?
>
> (TI, 127–128)

The risk involved in any such production of distance between the self and its elemental food is that the self would "lose the confirmation which, as living from . . . and enjoyment of . . ., it receives in the elemental, without receiving that confirmation from elsewhere" (TI, 128). The dwelling, transformed from a house to a home by the welcoming presence of the Other, offers a new source of nourishment, so that in retreating from its primal food the self takes up a new positioning rather than straying "into the void" (TI, 12). The energy driving representational distancing is drawn from a dimension which, though beyond sensory existence, is "lived positively [as nourishing] as a dimension of interiority starting from the intimate familiarity into which life plunges" (TI, 128). The intimacy of the home is essentially an intimacy *with some one.*

Yet the primary meaning of withdrawal into the home is economic rather than social:

> the first movement of economy is in fact egoistic – it is not transcendence, it is not expression. Work, which wrests things from the elements in which I bathe, discovers durable substances, but immediately

> suspends the independence of their durable being by acquiring them as movable goods, transportable into the house.
>
> (TI, 131)

Representation from the dwelling transforms the future as insecurity into the future as a "time-interval in which is inserted possession and work" (TI, 120), economic or acquisitive possessing takes over from the enjoying nomad's bountiful possessing-without-acquiring. So in this, its "first movement", representing from the dwelling is still in the service of the self's vital interests, rather than disinterestedly theoretical.

However, if the home to be the seat not just of discriminated and controlled substances, but of these substances as of objects of paradigmatic knowledge, there must be a dimension of dwelling which neutralizes the essentially appropriative character of being-at-home: there must be a mode of being-at-home which is being-*away*-from-home-at-home. There must be something eerie (*unheimlich*) about being-at-home.

And indeed Levinas tells us that in a second "movement", representing from the dwelling is not egoistic, but rather "liberates me from the very possessing which the welcome of the house institutes" (TI, 145), thus enabling me "to see things such as they are in themselves, that is to say, to represent them to myself, to reject both enjoyment and possessing" (TI, 145). For this, the mediation of a co-dweller is necessary. The Other, who makes dwelling possible by supplying the self's food once it is removed from the nourishing elements, also frees the self from the possessive pre-epistemic attitude it immediately adopts in the dwelling: the Other effects the self's transition within the dwelling from an economic to a knowledge-seeking stance.

More astonishingly still, this Other must be female:

> The welcome of the face . . . comes about originally in the gentleness of the feminine face, where the separated being may gather itself together and thanks to which it dwells, and accomplishes separation in its dwelling. Habitation and the intimacy of the dwelling which makes possible the separation of the human being, thus presuppose a first revelation of the Other.
>
> (TI, 124; see also TI, 128)

Levinas hastens to disclaim that he is "courting the absurdity . . . that every home presupposes *in fact* a woman" (TI, 130), a proposition which would entail that males unpartnered by females could not live in homes conducive to the separation (absolute posteriority of the subject *vis-à-vis* the world) required for paradigmatic knowledge. His idea is rather that it is "the feminine dimension" that is essential to the home. Whether the mere feminine dimension is capable of playing the role Levinas casts for it is a question I gladly leave to readers.

We might wonder what is special about *Woman*, and what is special about the presence of Woman *in the home?* Levinas's answer to the first question seems to be that woman packs a more powerful punch of alterity than Man: "[t]he Other *par excellence* is the feminine" (TI, 145), he says, reverting to a theme he had stressed in TA (77–81). The feminine is the very quality of difference rather than a set of properties different from those defining anything non-feminine. But is one female that different for another, or just for men? The answer to the second question is connected with the status of the home as the seat of representation of objects, thus potentially of paradigmatically known objects. The crucial role of Woman in the dwelling is to produce an away-from-home within the home: it is "through [the feminine Other that] *a world behind the scenes prolongs the world*" (TI, 145; emphasis added). Short of that 'prolongation' of 'here' into 'elsewhere', of at-home into away-from-home, dwelling would be a claustration, a perpetuation of imprisonment in (territorially expanded) selfhood. Woman is home-making in virtue of her capacity to domesticate the infinite, disseminate being-away-from-home within the home as its atmosphere of intimacy, "like a sweetness spreading over the face of things" (TI, 128). But the important point is that this sweetness is an absolute trace of a transcendence or infinitude beyond the 'digestive' capacity of the dwelling self. Woman incarnates the absent-presence structure of the apprehension of transcendence (absolute exteriority), for which

> *it is necessary that the presence of the Other* be revealed not just by the face which pierces its own plastic image, but also that it *reveal itself*, simultaneously with that presence, *in the Other's retreat and absence.* And the Other whose presence is discreetly an absence . . . is Woman
>
> (TI, 128; emphasis added)

Woman meets that second topographical condition through her language of silence in the home, her presence that is discreetly an absence (TI, 128), her "flight from the light" (TA, 79) in which the world of finite subjecthood is immersed. All this is well said by Browning in his *Love in a Life*:

> Room after room
> I hunt the house through we inhabit together
> Heart, fear nothing, for heart thou shalt find her
> Next time, herself! – not the trouble behind her
> Left in the curtain, the couch's perfume!

This woman is absolutely past, an original trace, never more than the trouble behind her; an intimation of transcendence (the infinite) to a subject suddenly removed away-from-home within the home, that is, still representing the world, only no longer from the standpoint of the topping self – in the ethical relation 'the topper is topped'. For the self to be affected by this atmosphere of trouble is for it to enter a dimension of 'untoppable' belatedness or absolute posteriority/passivity, originally glimpsed through Woman's absent presence in the home. In the home as a 'fortress' the world is represented as instrumental objects, but only in the home as co-inhabited by transcendence-disseminating Woman are objects represented as absolved from reference to a possessing self, such that "I situate myself above my engagement in the non-self" (TI, 145) and am able "to take an objective view" (TI, 145). The attitude of paradigmatic knowledge is thereby attained, scientific inquiry is instituted in a dimension of self-effacing (passive) exploration of absolute alterity, correlating with the knowing self's absolute posteriority in relation to its objects. Science on this train of reasoning is the investigation of the world from a standpoint elsewhere than the self, from the standpoint of "the Other within the Same"; or of the subject conceived as a passive "subjectum" (AEE, 147). Man is the author of science, yet ultimately not as the mastering Baconian subject; for despite all its active theoretical construction, the 'ontological' *dimension* of the scientific endeavour is ethical (as distinct from causal) passivity in the face of the absolutely-other-than-me that rests on the absolutely-before-me.

Subjectivism?

'Taking an objective view of things' means, for Levinas, representing them as what he calls "things in themselves". His usage of 'thing-in-itself' differs from the traditional (Kantian) one in that it does not refer to a status material entities would possess in the absence of cognitive subjects. For Levinas as for all phenomenologists, the concept of thing-in-itself implies subjectivity. Husserl, Heidegger and Levinas differ only in their versions of this subject-boundness of in-itselfhood. All reject "the concept of transcendence [as] the absurd positing of a being behind the scenes in a world behind the scenes".[15]

According to Husserl, the in-itselfhood of a thing is a fact about its transcendent existence *for consciousness*: no consciousness, no sense in saying that a thing exists in-itself or independently of consciousness. Levinas agrees that "we ought not to require for things an existence independent of consciousness, for their independence is not, on our view, to be construed as a negation of transcendent existence but as a characteristic of it" (TIPH, 80). But note that the transcendence in question in this context (a discussion of Husserl) is objectivity constituted by intentional consciousness, not Levinas's *pre*-intentional transcendence (absolute exteriority). The distinction is structural to the whole of TI, as its Preface warns: "[t]he difference between objectivity and transcendence will serve as a general guideline to all the analyses of this work" (TI, 5). Levinas's agreement with Husserl is therefore subject to the proviso that the latter's intentional objectivity is not to be understood as the *original* manifestation of transcendence.[16] Husserl's objectivities refer *back* to consciousness as to their constituting ground; Levinas's transcendent objects exist as absolving themselves *from* relation to consciousness. In neither case, however, is it true that the object stands in absolutely *no* relation to consciousness.

But that difference in the interpretation of the subject-boundness of in-itselfhood is a family quarrel compared to the difference between accepting and rejecting the principle that in-itselfhood is in *any* sense bound to subjectivity. Conventionally, in-itselfhood is construed as something things would still enjoy even if all human beings were to vanish from the face of the earth. If this absolutist conception of in-itselfhood is right, the whole principle of the phenomenological notion of subject-bound in-itselfhood must be wrong. It must be untenable subjectivism.

To the charge of subjectivism thus laid, Heiddegger offered an explicit answer which is worth considering before seeing whether Levinas can benefit from the same sort of defence. "Of course, only as long as Dasein *is* (that is, only as long as an understanding of Being is ontically possible) 'is there' [*'gibt es'*] Being."[17] Heidegger is not saying here that the entire physical universe would dissolve into nothingness if Dasein were to vanish from it. That this is not what he means is made clear in the immediately preceding sentence: "[the proposition] that reality [*Realität*] is ontologically grounded in the Being of Dasein cannot mean that what-is-real [*Reales*] could only be, as that which it is in itself, if and so as long as Dasein exists".[18] These two statements are not mutually contradictory: in a world from which Dasein had vanished, what-is-real (*das Reale*: material and psychological entities) would not have vanished, for Heidegger does not claim that Dasein creates matter or states of mind: "the characterized dependency of Being, *not of entities*, on [Dasein's] understanding-of-Being, that is to say, the dependency of reality, *not of what-is-real*, on concern . . . etc."[19]

What Heidegger positively claims is that, were Dasein to vanish, then along with the understanding-of-Being unique to it would vanish the *that there is* . . . of entities and their status *as* exemplars of modes of Being (e.g. 'reality', 'things-in-themselves'). It would cease to be the case *that there are* trees, though this would not affect the habits of the myriad of organisms that would 'continue' to live from them. In the conjectured circumstance, it would not be the case that trees would still *be*, only without anyone around to *say* that they are. For *there being* things (and there not being things) requires Dasein's understanding-of-Being and, on a par with this, its revelatory (as distinct from recording) use of speech.

Now, that there is a mode of Being 'independence-from-Dasein' is part of Dasein's understanding-of-Being. But with Dasein quitting the scene, independence-from-Dasein would also quit it: "if Dasein does not exist, then 'independence' 'is not' either, nor is the 'in-itself' ".[20] Essentially the same defence could be mounted by Levinas, making the requisite substitutions: 'if the time of the Other does not exist, then independence-from-self is not, nor is the in-itself'.

Part II

Confrontations

4 The language of time

The constellation of positions

In this chapter I shall be concerned with responses I believe phenomenology of time is able to make to challenges issuing from the language-based strand of analytical philosophy of time. To what extent phenomenology of time could answer potential criticisms from the science-based strand will be a matter for Chapter 6. These two strands are substantially separate, inasmuch as the first focuses on interpretations of time in the various branches of physical theory and the second on problems latent in our common-sense concept of time, problems which would arise even if physics did not exist.

Contrary to popular belief, the idea at the heart of Bergson's philosophy of time is not opposition to spatializing time, a vice also abjured by cohorts of philosophers with no liking for Bergson or phenomenology, but rather the claim (considered above, p. 12) that the flow or passage of time implies no thing that flows. When a thing does happen to be involved, human *durée* changes even if no change is discerned in the thing:

> for all that the object remains the same, for all that I may look at it from the same side, from the same angle and in the same light, my vision of it now nevertheless differs from that I have just had, if only for the reason that I have aged a little. My memory is there, which pushes something of my past into my present.[1]

This visual experience changes in something like the way a musical note does when held without alteration of loudness, pitch or timbre. Time passes even when 'nothing happens'. Husserl claims the same autonomy of time-change for his time-consciousness: "any object that changes is missing here; and since 'something' runs its course in ever nothing here that changes, and for that reason it also makes no sense to speak of something that persists"[2]. This time is not dependent on (the perception of) change-events befalling things, given that in it no thing flows. Sartre too signs up to the autonomy thesis in declaring that "pure and absolute change . . . can, moreover, perfectly well be change without there being *anything* that changes, and [this] is *durée* itself".[3] Also Merleau-Ponty: "What happens [*se passe*] in time is the passage of time itself".[4]

On this evidence it might seem that the autonomy thesis entails, if it is not identical with, the thesis of time-motion in the sense of the passage of a *durée* in such a manner as to conserve itself in its totality. What is clear is that an entailment holds in the opposite direction; time's passage entails its independence from the change of things and events, since otherwise it would be the passing-*away* of time. Yet time does not pass away in passing; it conserves itself (its past phases) as a whole, as the 'durational' doctrine points out. Conversely, things and events do pass away, do not conserve themselves in their totality (or at all). Passing-away is what befalls, say, the wholly green state of an apple when the apple turns red all over, since the phenomenal red apple does not conserve the phenomenal green apple.

As to the question, left open, of the reverse entailment: Does the autonomy-thesis entail (durationally) passing time? I see no reason why it should; and indeed the actual constellation of positions held by phenomenologists suggests that it does not. For both Heidegger and Levinas espouse the autonomy thesis while denying that fundamental time is passing *dureé*. When Levinas says, "[t]ime itself refuses all hypostasis [reification], the images of flow and flux by which people explain it apply to beings in time and not to time itself. Time does not flow like a river,"[5] he is endorsing the autonomy thesis ("time itself") and rejecting the durational thesis ("does not flow").

A further point worth making in relation to the constellation of positions is that in point of historical fact the durational thesis – essentially the idea that time conserves itself in passing – is not identical with Bergson's inaugurative version of it. Husserl, Sartre and Merleau-Ponty, though appropriating the kernal features of self-conserving passage and autonomy, reject the specifically Bergsonian determination of *durée* as a *creative* conservation not radically distinguished from a *natural* process.

The notion that time passes has, however, been severely censured, indeed ridiculed, by the major positions in analytical philosophy of time. If this has not taken the form of an explicit attack on the phenomenological-durational version of time-motion, it is only because the assailants studiously ignored the existence of the phenomenology of time. The target has been offbeat non-phenomenological theories of transience and our philosophically unarticulated everyday locutions: *time passes, time flows*, etc. But since I am certain that this criticism would look no more kindly on the phenomenological-durational version of the passage of time, I shall assume that phenomenology is in the position of having to defend itself.

In the first section of this chapter I shall defend what I understand to be the Husserlian model of the passage of time (endorsed by Sartre and at least the early Merleau-Ponty). In that the time it construes as passing is serial time, Husserl's model provides a significant basis for confrontation with the exclusively serial conceptions of time proposed by analytical philosophy; whereas Bergson's time would yield a less significant confrontation owing to the fact that it does not pass in serial style (Bergson denies temporal order outright: see below p. 181). My defence of Husserl's passing serial time will argue that it is secure against the objections levelled at time-motion by analytical philosophers, reserving judgment as to whether it is fundamental time.

The objection to 'time passes' is coupled in the minds of those who raise it with the folly of treating time as some sort of dynamic substance which gives rise to pseudo-perplexities and regresses. Having answered this charge in Section 1, my aim in a complementary Section 2 will be to argue that perplexity over time arises rather from the very *un*thinglike nature of passing time – its failure *to be* (present), its lack of substantiality. Since notions of substantialization of time have had currency in analytical philosophy in recent decades,[6] we would do well to be clear about how they differ from the substantialization connoted by the autonomy thesis. To this end I shall consider, first, the version of illicit substantialization imputed by the tensed theory of time to the tenseless theory, and, second, the version imputed by the tenseless theory to the tensed theory. Preliminary to this, however, some clarification of the differences between these two 'analytical' theories is in order (this will also pave the way to the considerations of Chapter 5).

That there are temporal relations of the type 'A is earlier/later than B' is germane to our ordinary understanding of time. The events that are the relata of these relations cannot change their intrinsic properties or their temporal position: event A cannot start off by being earlier than event B and end up later than it, or even closer to it than it was. It is also clear that our ordinary understanding of time involves more than these time-relations. For the bare information that 'A earlier than B' does not tell us how that relation stands with respect to now: now might coincide with B, in which case A is past, as well as earlier than B; or it might coincide with A, in which case B is future, as well as later than B. The same events enter into two series, the one called 'tensed' because it is comprised of events as changing in relation to now (from future to now or from now to past), and the other 'tenseless' because comprised of the same events as earlier or later than each other irrespective of how they stand to now. The Battle of Waterloo stands in the same invariant relation to the Battle of Hastings today (now) as it did a century ago, when today's now was far in the future, and as it will in the year 3000, when today's now will be far in the past. That time involves these two types of temporal relations is accepted all round.

Contention enters in with the claim that one of the two types of relation is conceptually more basic than the other. Common sense holds that the tensed type is more basic (explanatory) than the tensed: only once the two mentioned battles have come to pass does it become the case that the one is thereafter invariantly later than the other. This is the view defended by tensed theorists. Tenseless theorists, on the other hand, claim that the tenseless relation is basic and, as such, deserving of the title 'real time', while tensed time is relegated to a subjective perspective on tenseless time. This means that the tenseless relation 'A is earlier than B' does not, contrary to common belief, presuppose that A *came to pass* before B (that A is past when B is present). Nothing, in fact, comes to pass or passes away on this interpretation of tenseless time, which means that the membership of the tenseless series can neither increase nor decrease, and that the total quantum of evil that has been and ever will be perpetrated impinges with full force at any now. It is this denial of 'becoming' that sets tenseless theory apart from tensed theory. For if the events A and B were obliged to come to pass in order for the tenseless relation 'A is earlier than B' to hold, all the times at which both of them have not yet come to pass would be times when that relation does not yet hold, which is to say that it would not hold at all times. To feel comfortable with the tenseless theory is to have no difficulty with the notion that "[e]vents do not happen; they are just there and we come across them".[7] It is not easy to accept this invitation to sever the concept of an event from the concept of a happening, or convincing to be told that events are indeed happenings, only not such as involve coming-to-pass. I find myself thinking of a constellation of eternal stars valiantly twinkling for eventhood, though I doubt the image eases the strain the tenseless theory puts on our concept of event.

It is to this difficulty of prising off the concept of an event from the concept of becoming (present and past) that the tensed theory ultimately appeals in bringing its charge of substantialization of events (time) against tenseless theory. Events are timeless things in a quasi-space if in order to be they have no need to come to pass and pass away. Or as the tensed theorist Richard Gale put it: "physical events are substantialized because they *co-exist* in a one-dimensional *spatial* order and *endure* in the same spatial order".[8] (The tenseless theorist replies that tenseless events 'co-exist' only in the sense that they do not become, not in the sense that they are not successive, and that their becomingless *yet successive* order is sufficient to exclude substantialization/spatialization. But of course 'becomingless yet successive' is the bone of contention.)

The notion of tensed time as a subjective perspective on this allegedly pseudo-temporal order was once well captured by C.D. Broad in his metaphor of a policeman's torch playing along a line of doors, its light falling now on one door and now on the next. The spotlight stands for the subjective perspective 'now'. All the doors already co-exist in a one-dimensional spatial order no matter where the spotlight falls. A given door becomes present when the spotlight plays over it, past when it moves away. But its coming under the 'spotlight of the now' is not its entry into reality, nor does the withdrawal of the spotlight consign it to unreality.

Let us now see how the tenseless theorist returns the compliment. Consider the fact of a traffic light changing from red to amber. Both theories agree that the traffic light (a thing) changes, and that its changing consists in an event happening to it. The tenseless theorist maintains that this event or happening in no sense changes. Events *are* changes and therefore do not change. Events are the unchanging changes that time is, and expecting them to change would be treating time as a substance which changes its properties. The tensed theorist agrees that the change from red to amber is an event befalling the traffic light, such that it (a thing) changes its intrinsic properties or relations, but he or she claims additionally that the change-event, though not changing its sensible properties, does change its tensed determinations: changes, that is, from being future to being present to being past. But here the tenseless theorist objects that this is to treat the events of which time is comprised as timeless things or *substances* which don and doff 'temporal properties' (futurity, presentness and pastness).

It is evident that neither of these alleged substantializations of time is anything like as radical as the phenomenological substantialization that I have been calling the autonomy thesis. For in both cases time remains bound to an order of change-events and the alleged substantialization of time consists in subjecting these to the logic of things. Time is not severed from events and considered as something independent, but rather events are treated as timeless substances, spatialized or otherwise eternalized.

The fact that tensed time might be vaguely described as a dynamic time suggests that it will give us more immediate access than the tenseless theory to comparisons with phenomenological dynamic time (passing time). So it will be worthwhile deepening our understanding of the tensed theory before attempting to pin down just how it stands in relation to the phenomenological conception of the passage (dynamism) of time.

We may begin by noting that the tensed theorist can reply to the tenseless theorist's charge of substantialization that it relies on a false analogy between events becoming present (and past) and a substance undergoing change; between what Gale, following Broad, has called "the change *of* time" as distinct from "change *in* time" (LT, 242). The former is 'time itself', conceived unphenomenologically as the change of tense *of events*, the latter is the change of properties undergone by the things to which events happen. The change of time (an event's becoming present and past) is presupposed by change in time: a thing changes because it is hit by an event – but why wasn't it hit five minutes ago? Because five minutes ago the event had not undergone the change from not becoming present to becoming present.[9] The tensed theorist can press the point that the non-sensible (purely temporal) changing of an event from future to present to past does not cast the change-event in the role of a timeless substance; for a substance exists before its changes of property, whereas a change-event does not exist before it becomes present; and a substance survives the change-event befalling it, whereas a change-event does not survive becoming past: the traffic light survives its change of colour, but the change of colour does not survive becoming past.

This defence by tensed theory might seem to bite its own tail, however. For in insisting that an event does not exist before it becomes present, it ties the concept of an event so tightly to the concept of becoming-present that we might be inclined to question whether we are really adding anything to the concept of an event by saying that it becomes present (and past). Is Gale's change of time not just so much excess (and potentially obfuscating) metaphysical baggage? Let us try getting rid of it by saying that 'is past' applies, not just to event *e* but to the whole fact of *e*'s presentness. Add to this the stipulation that there are no truths (facts) about what is not currently in existence. Then 'e is past' means that '*e* is present' was but no longer is a present truth, and there are no non-present truths about *e* (such as that it 'exists in the past'). But let it not be asked, 'How does that change of truth-value come about?' For the answer could only be 'because of becoming'. This position voluntarily blinds itself to the fact that the essence of tensed time lies in the *shift* (change of time) which the change of truth-value merely records in a way that is wholly uninformative about what it is for an event to change from present to past. The change of truth-value may be a criterion for an event being in tensed time (the time in which events 'come and go'), it is not a definition.[10] (Much the same may be said of Russell's tenseless 'definition' of change: see below, pp. 155–156. Analytical philosophy of time has been fruitful in the invention of devices for screening out the actual 'event' of change.)

The change of time is Gale's construal of 'time passes'. The expression is for him legitimate provided it is taken as referring to no more than the change of events from future to present to past. So conceived, the passage of time does not depict events as subject to the temporal logic of things (substances); we do not treat the change of time as happening *in* time. Of events we can legitimately say that they 'come and go', and call that 'time passing' if we like. We can say, on Gale's showing, that events become present and past, but we cannot legitimately say that "new events are *always* becoming present" or equivalently, that "time is *continually* or *always* passing" (Gale's examples: LT, 242). For in these popular aphorisms the adverbs express temporal concepts which properly apply only to the change of things in time. In applying them to the change of time, we turn this into a thing which changes (in ways described by adverbs), we substantialize events (time), construct a pseudo-concept of the passage of time involving a vicious infinite regress of meta-times. *Continually* properly refers to how *things* are in a state, not to a way the events befalling them have of becoming present and past. Likewise, the adverbial concept of *rate* applies only to the change of things, as evidenced by the appropriateness of asking how fast a traffic light changes from red to amber as against the absurdity of asking how fast that change-event becomes present. So the general message from Gale is: "What time makes it possible for us to say is exactly what cannot be said about time" (LT, 243).

Unlike many philosophers (including many tensed theorists), I believe there is substance in this Gale–Broad distinction between the change of time and change in time. As undergoing the change of time, things change absolutely, whereas their changes in time permit of degrees: when a poker becomes less hot its former hotter state is not concomitantly *less* present than it was, nor its initial cold state a lot less present – these two past states (hotter and cold) are *equally* non-present. On the other hand, now that I am conscious of a less hot poker, my *consciousness of* the hotter poker *is* less present than it was, and my consciousness of the cold poker a lot less present – that's where there's room for the passage of time *as something distinct from the passage of events.*

The Gale–Broad distinction between the change of time and change in time has interested us because it gestures towards phenomenology by introducing time-levels and consequential constraints on the use of time-language across levels. A general principle endorsed by Husserl and Gale–Broad is that it is illicit to speak of a more fundamental time-level in the language of a less fundamental one analysed in terms of it. But there is this important difference between the application of the principle in the two cases: though *no language* of time-consciousness in fact exists that does not prey upon the language of a less fundamental event-bound time, Husserl can in principle speak about the becoming of events – describe how they become individuated – just because his time-consciousness is more fundamental than the becoming of events, whereas Gale must enjoin silence as to the becoming of events, since he acknowledges no *time* more fundamental than it. The only other time he acknowledges, change in time, is less fundamental.

Husserl's wrapping-up in time-consciousness of Gale's change of time suggests it might be worth going back to those aphorisms. Are they really meaningless, incoherent, despite countless generations of people having believed them true? Must we interpret the mentions of *always* and *continually* as generating absurdities and vicious infinite regress? While it is true that the aphorisms are semantically equipped to sustain Gale's regressive analysis, this is not proof that they could not be given a different analysis. Given the indeterminacy of language, and especially of contracted aphoristic language, it seems reasonable to explore that possibility that, while the adverbs do allude to an enveloping time, this need not be construed as a regress-generating reduplication of change in time. Might the time that envelops the change of time in Gale's sense not be a wholly different sort of time – an unenveloped enveloping time that passes in virtue of its intrinsic event-free structure? The arguments of the first section will press the case that this sort of passing time is in fact supplied by Husserl's concept of absolute time-consciousness.

The pertinence of the supposition of an unenveloped enveloping (thus absolute, non-regressive) time-order is strengthened by the consideration that, if events are not allowed to become present (continually) in such a time, then it will not be possible to say *when* (at which now) they become present. But surely events do become present (and past) at nows (even if not at rates), despite this being a case of applying a temporal concept (*when*) to time (events). That we have a problem here of 'saying when' that is not removed by Gale's ban on applying temporal concepts to time becomes even more apparent when we realize that the regress breaks out just as virulently in the case of the approved application of the temporal concept *when* to the change of genuine *things* in time. 'When does the traffic light change from red to amber?' The relations of tensed series do not supply an answer: it's no good saying that the traffic light changes from red to amber at a now that is now when its changing from green to red is just-past; for there is no way of picking out this particular case of just-pastness from all the cases there are without presupposing as fixed the identity of the now in question. All we have said is that when*ever* the one change is now, the other is just-past. Whence the sirenic lure of the meta-series as picking out, from all the nows there are, the one at which a change happens; or the unlovely lure of tenseless time, assuring us that it is at *all* tenseless dates the case that the succession 'A just earlier than B' holds.

Section 1 – In defence of passing time

Tenseless theorists commonly assert that any dynamic theory of time is a mistake arising from "our ordinary-language expressions", from the temporal metaphors which "continue to confuse us about the nature of time".[11] As against this, I want to argue two claims:

(A) That (a) some interpretation of the metaphors' apriori can be hermeneutically true of our core understanding of time; and (b) that apriori only one interpretation is admissible.
(B) The one admissible interpretation is substantially provided by the Husserlian time-model.

(A)

(a) It is only when on intellectual holiday that tensed theorists would say that time is a 'flow of events'. Back in the study they say that 'Tensed sentence-tokens have irreducibly tensed truth conditions.' The tensed theorist can dispense with the temporal metaphors, whereas the phenomenologist is stuck with them. Phenomenological theories of passing eventless time are ineffable without the metaphors. But this means that the truth of these theories would imply that there must be such a thing as the truth of metaphor and that any defence of phenomenological time-motion would be a defence of the cognitive significance of a metaphor. This would be taking metaphor seriously, as something which can be hermeneutically true in the sense of clarifying understanding.

It has been standard practice in philosophy to pre-ordain the incapacity of the temporal metaphors to be cognitively significant by assuming without argument that cases of taking them seriously could only be cases of taking them literally, of assenting to literal falsehoods: "[dynamists] take the river-of-time model seriously . . . the adherent of the dynamic view of time takes these expressions not really as metaphors at all. They express for him the honest truth about reality. Time flows."[12] It is hard to see how this writer can believe it psychologically possible for even philosophers to suffer from the delusion that time is literally a river in which they might drown. We may presume that nothing so bizarre is meant, but also that nothing significant has been said. Taking metaphors literally is only a fool's way of taking them seriously. The sane way involves rejecting the idea that because they are fictions they must be without hermeneutical significance (powerless to disclose reality), at best 'mere' and at worst treacherous. Considering the hermeneutical roles metaphor is now acknowledged to play in aesthetics, in the natural and human sciences, what is required in place of the outdated dogma of 'decorative and dangerous' metaphor (to which philosophers still resort wherever they think they can get away with it) is an exploration of the possibility of metaphorical truth and an understanding of how it is possible for us to break the rules of literal truth-getting.

Yet there might seem to be grounds for supposing that the specifically temporal metaphors could not share in the cognitive upgrading from which metaphor has benefited in recent decades. These grounds were once elegantly stated by Broad as follows:

> [temporal metaphors] give no help towards analysing or comprehending generic facts about time. A metaphor helps us to understand a fact only when it brings out an analogy with a fact of a *different* kind, which we already understand. When a generic fact can be described only by metaphors drawn from specific instances of itself it is a sign that the fact is unique and peculiar, like the fact of temporal succession and the change of events from futurity through presentness to pastness.[13]

Broad's first point here is that the temporal metaphors are a peculiar species, in that they are of the form 'A is a^1, a^2 . . .' (drawn from the generic fact to which they refer) and not of the helpful form 'A is B', which is the usual form, exemplified by the proverbial 'Man is a wolf'. It conveniently follows that my purpose here would not be served by trawling the vast literature on metaphor of the usual 'A is B' form for theories congenial to the proposition that metaphors of that form can be hermeneutically true (though support could be got from the interactionist theory). For the temporal metaphors are not of that form. Broad's second point is that the fact that the temporal metaphors are of the form 'A is a^1, a^2 . . .' means that they are cognitively empty – "give no help". The generic facts in question are temporal facts (succession, etc.). A specific instance is a "flowing stream" (Broad). But since this is a temporal process, it contains everything that is obscure about the generic temporal fact it is called upon to illuminate. The case is different with 'Man is a wolf' since a wolf is (i) a different kind of thing from a man; and (ii) a thing we already understand. I must take issue with this argument since it would scotch my proposition that on a certain interpretation the river metaphor is hermeneutically true of the core of our understanding of time.

Broad's suggestion of *cognitively vacuous* circularity derives from his tacit premiss that understanding specific facts presupposes understanding the corresponding generic facts. But this is only a half-truth, the other half being that it is *also* the case that understanding generic facts presupposes understanding the corresponding specific facts. Or at least this is the case from a phenomenological standpoint where 'generic fact' means 'phenomenological essence' and 'specific fact' means 'example of a phenomenological essence'. Granting these glossings, let us check out this two-way presupposition in the domain of the imaginary, before returning to time.

Consider yourself contemplating a mental image. You know that the image exemplifies the essence of the imaginary and that *in meaning* it is univocally prior to its examples (in the same sense as the meaning of triangularity is not exhausted by any particular triangle). But the issue is whether the imaginary is univocally prior *in knowledge* to its examples, since Broad's claim is that we get to know nothing about the essence 'time-flow' from the example 'flowing river'. Now it might seem as if our ability to recognize particular experiences as (e.g. mental) images attests to the essence of the imaginary being univocally prior in knowledge, for do such *re*cognitions not presuppose that I am already acquainted with the imaginary? But then it is very fishy that we never catch ourselves understanding essences directly, without going via their examples. It is clear that the only way we can get to know about the essence of the imaginary is by inspecting concrete images. Yet it is equally clear that recognizing these as such (and not as, say, perceptions) presupposes prior knowledge of the essence of the imaginary. There is nothing scandalous about this circle. It is the benign circle of hermeneutical understanding, and it is cognitively productive rather than vacuous since the examples enable us to raise to conceptual clarity what we knew only in an implicit pre-understanding. Broad's position seems to be that the essence's univocal priority in meaning entails its univocal priority in knowledge, in which case it would indeed be true that from the examples we could learn nothing about the essence.

So there is no apriori reason why temporal processes could not teach us about time, *provided that* in observing these processes we implement Husserl's method of "free variation in imagination", through which the essential features of an experience are shelled out of the husk of non-essential features in which they are empirically embodied. Wetness and pools, for example, would be identified by this variation as non-essential to the concept of time-flow, as would also the fact that empirical rivers have sources and estuaries (whereas time has no beginning or end). The fact that human beings have always looked to rivers (to metaphors of the form 'A is a^1, a^2 . . .') for lights on time is just as it should be.

Broad's approach to temporal metaphor is interesting because it implicitly acknowledges the possibility of metaphors being hermeneutically true ("helpful") whereas the earlier mentioned approach dismisses them on the ground of an antedeluvian category mistake. In saying that the truth of the temporal metaphors could only be hermeneutical, I mean that they would be revelatory of the nature of time and therefore a different sort of truth from correspondence truth; the correspondence truth of the statement that there is a cow in the field presumably does not enhance our understanding of either cows or fields. While this is no place to attempt to go into the differences between the two kinds of truth, it seems evident that they have in common the concept of 'fitting reality'. If the temporal metaphors can be true, it must be thanks to some sort of 'fit' between the structure of the specific fact (river) and the structure of generic fact (time), such that the latter is rendered visible in the former. But what sort of fit?

Metaphors of the form 'A is B' involve a disruption of two normally disjunct sets of sub-categorization rules: saying that man is a wolf disrupts man-language as much as wolf-language. By contrast, saying that time is a river disrupts only our literal river-language, not also a literal language of time. For we have no literal language of time. Time is spoken of in the language of space and of physical motions in space. In other words, the language of time *began* as metaphorical, it is absolute metaphor. Even 'present', and therefore 'past' and 'future' too, is infected with the 'here' of tensed space. I now want to suggest the possibility of grounding the truth-capacity of the metaphors of flow on the good fit (congruence) between the originally metaphorical or shifted nature of the language of time and the originally metaphorical or shifted nature of time itself.

We find, among others, time-theories in which:

(i) Time is thought from presentness.
(ii) Time is not thought from presentness.
(iii) Time is thought from the absence of presentness.

(i) and (ii) are exemplified by tensed and tenseless theory respectively. Husserl's theory comes under (i) inasmuch as "[w]hat is actually present now is there itself. And what is there itself individually is present now" (ZB, 211). The self-temporalization of time from the central now coincides with the self-constitution of 'senses', including the sense 'truth'. Derrida accurately records Husserl's position when he writes: "the present now governs every possible concept of truth and sense".[14] This sort of sense, constituted by consciousness, contrasts with Heidegger's "sense of Being", which is at the heart of time-theories of type (iii). A remark by Levinas should give us some purchase on this: "being is what it is, what it shows itself to be in its truth and, at the same time, it resembles itself, it is its own image. The original . . . withdraws, as if something in being lagged behind being".[15] Something unpresented haunts perfectly visible things, *especially* when they innocently declare themselves to be just as they appear. As we have seen (pp. 64–65), a thing *is* not its appearance, it 'is' its appear*ing*. In the wake of its appearing (its) Being withdraws, hides, differentiates itself from appearings, asserts its *ontological* difference from them. What is happening here is that in becoming present something yields a sense of Being, leaves a trace of its absconded ground. The present is the making-present or presencing of something having that sense. But now consider: the passage of time is the transfer of sense, meta-phor: presentness^ transfers itself in durational consciousness from future presentness to past presentness to past past presentness. The past is future presentness become past presentness. Pastness is a *grounded* metaphor of presentness in that as an eidetic sense presentness is prior to it. But time as the transfer of *this* sense of presentness (as *durée*) happens in the key of an absolute pastness or "prehistory",[16] within a dimension of absolute (groundless) metaphor, inasmuch as the presentness that gets transferred also, more fundamentally, has sense as (non-eidetic) sense of absconded Being. Time as the transfer of the eidetic sense 'presentness' conjugates, so to speak, that infinitive sense of Being. Congruently, the language of time is a stock of semantically grounded metaphors preceded by no literal language. As time-metaphor stands to Being, so the linguistic metaphors of time stand to literal language of time. I believe this congruence answers the question of the apriori possibility of the temporal metaphors being hermeneutically true of time.

(b) Why can there be only one admissible interpretation of the temporal metaphors, immemorial and unrevisable whatever it is? The reason is implicit in the explanation just offered for the truth-capacity of the temporal metaphors. Any expression whose meaning is revisable belongs to the ordinary (the linguist's) hierarchical time of language, in which literalness logically and historically precedes metaphoricity, metaphors 'die' and are reborn as literal expressions, historical interpretations of metaphors give way to reinterpretations. But the absolute nature of the metaphors of time shows that they do not belong to this ordinary time of language in which linguistic meaning undergoes re-interpretation. They are therefore exempt from the vicissitudes of re-interpretation that befall expressions born in the time of language which begins with literalness.

Under (A) I have argued (*a*) that some interpretation of the flow metaphors apriori can be true and (*b*) that only one interpretation is admissible. Obviously I cannot now proceed by inferring the sole admissible interpretation from the theory of time I claim the metaphors are true of, since other interpretations, *if admissible*, might fit other theories just as well. So my immediate task must be to establish theory-independent criteria of admissibility. This done, I shall try to show that the only interpretation admitted by the criteria fits the conception of time-consciousness that Husserl dubs "a flow".

(B)

The first criterion is supplied by our principle that re-interpretation of the metaphors is inadmissible. Consider the following would-be sanitized usage recommended for 'time flows' by a scientifically-minded tenseless theorist who wants to purge the metaphor of the suggestion that real time involves tensed becoming:

> time 'flows' only in the sense in which a line flows or a landscape 'recedes into the distance'. That is, it is an ordered extension. And each of us proceeds through time only as a fence proceeds across a farm: that is, parts of our being occupy successive instants and points respectively. There is passage but it is nothing extra. It is the mere [*sic*] happening of things, strung along the manifold.[17]

This is what Gale disparagingly calls "the frozen River of Time" (LT, 230) metaphor, according to which the tensing human mind crawls along its world-line, bestowing the property of presentness on the objectively tenseless events it comes across. This obviously revised scientific interpretation of the metaphor cannot be legitimate if I am right that the meaning of the flow-metaphors is not up for grabs. The only proper use a tenseless theorist can make of these metaphors is none. (My position here differs from Gale's as follows. Without considering alternative interpretations, Gale pronounces the metaphor guilty by association with a theory he believes false, whereas I bracket the truth-value of the tenseless theory and deny that the metaphor may be legitimately re-interpreted to fit it.)

A second criterion is that no interpretation of the river metaphor is admissible that renders the river redundant. This criterion invalidates proposed interpretations which, unlike the scientific-tenseless case just considered, are not *obviously* re-interpretations, in that they retain the commonplace association of the metaphor with dependent motion, or flow *in* the river. I suspect that many people would agree with A.N. Prior that the following, for example, correctly explicates 'the' or 'our' pre-scientific intuition that time flows:

> it was the case that *p* but it is not now the case that *p* – this formula continues to express what is common to the literal flow of a river on the one hand (where it was the case that such-and-such drops were at a certain place and this is the case no longer) and the flow of time on the other.[18]

Here the river metaphor is being pressed into the service of a view of tensed change which, as I shall argue later (pp. 151–152), is unacceptably atomistic. The point to note for the moment is just that Prior's formula would apply no differently if for discrete drops were substituted logs or even objects whose motion is not conditioned by the motion of a river at all. In other words, this analysis of the river of time renders the river redundant.

But perhaps this is doing Prior an injustice. Since he believes that time is wholly comprised of a series of tensed events it might seem only charitable to suppose that he chose "such-and-such *drops*" (rather than logs) as his value of *p* precisely because rivers are physically comprised of drops. Does the fact that a river is *made up* of drops not mean that in analysing the river metaphor in terms of drops we are *not* rendering the river redundant? Taking Prior this way, his position is more clearly stated by another philosopher as follows:

> events do not just have their places in time like pieces of wood floating in a river, but events *constitute* time . . . *the flow of time is nothing but the flow of events*. Therefore we should not compare events in the flow of time with objects floating in a river, but with the molecules of water the river is composed of. As the passing of molecules of water constitutes the flowing river, so the passing of events i.e. their occurrence, constitutes the flow of time.[19]

The point this writer wants to make is that the order ("flow") of time is generated by the occurrence of events themselves: events do not fit into a pre-established time-order, like logs following the flow of a river: there is no flow of the river distinct from flow in the river – "nothing extra". This gives us a working river, though one comprised of molecular motions no one has ever seen. The question is whether this could be a river into which the metaphor could be legitimately cashed. That it could not may, I think, be brought out in the following way. Supposing it were true that unperceived objects are colourless, it might follow that we could not truly say that monsters have green eyes, but not that there can be no truth in the metaphor that jealousy is a green-eyed monster. For one thing the 'seeing' that jealousy is a green-eyed monster requires is just that monsters be seen to have green eyes, not that their eyes be really green. Metaphor happens in the domain of the sensory. So, *just because it is a metaphor*, 'the river of time' is tied to a meaning offered in the contemplation of *visible* rivers. If we choose to reconstruct rivers scientifically, in terms of invisible molecules, then we have installed ourselves in the non-sensory domain of the invisible and forfeited the right to invoke the ordinary pre-scientific meanings of words like 'flow' and 'passage'. Yet the enterprise of corrective re-interpretation depends for its semblance of viability on poaching descriptive terms from the pre-scientific language of the visible; on using, as in the quote, words like "passing" and "flow" in a sense that is at once shifted (since they did not before refer to an invisible molecular flow) and not shifted (since the recommendation is that we rethink the flow as molecular flow). Straight speaking means opting for the one language or the other, not one for the science and bits of the other thrown in to show how science corrects our everyday notions. (None of this prejudices the issue of the commensurability of these two languages – my point is just that commensuration is not helped by mixing.)

A third criterion is that no interpretation is admissible if it entails absurdities. Repeatedly, analytical philosophers have claimed to reduce to absurdity all talk of time passing, though I now propose to show that their objections depend on caricaturing time-motion as a species of atomistic physical motion. The objections to the passage of time were once succinctly summarized as follows:

> 'the flow of time', 'the course of time', 'the passage of time' are all spatial metaphors carrying the same implication – that of some mysterious kind of fluid that 'passes'. 'Passes what?', we may ask. The absurdity of this idea is emphasised when we speak of time passing slowly or quickly, for how can we measure the pace of time? The measure of speed can only be in terms of distance in space, covered in a certain period of time taken as a unit. Speed, therefore implies space as well as time, so that to talk of time as a movement (flow, passage or what you will) or of the pace at which it passes is not only an example of spatialization, but also a descent into an infinite regress, because a presumed *movement* of time requires another time in which to move. That ulterior time-series will then also be a 'passage', and the infinite regress is evident.[20]

There are three objections here. The first has it that the unavailability of an answer to the question 'Passes what?' implies that the notion of time passing is absurd. To which it must be answered that it is the question itself that is absurd, in that it attempts to press time-motion into the conceptual scheme applicable to moving objects which pass by a point of observation external to their trajectories. Put differently, the question is blind to the fact that when we see rivers as images of passing time, it is the *internal* structure of the flow itself that we have in mind, not the changing relation of river-borne objects to an extratemporal observer standing on the bank. The *symbolically relevant* river changes *in relation to itself* (let's say for the moment). The perceptual feature of a flowing river that qualifies it as a privileged image of time is surely the simultaneously self-dissociative and self-re-integrative structure of the chasing of wave by wave. As Sartre says in a rough approximation: "psychologists who first noted the characteristics of the *durée* of consciousness very often compared it to a river. A river best evokes the constant interpenetration of the parts by the whole and their perpetual dissociation and free movement".[21] Or as Merleau-Ponty puts it in an expanded description:

> we say that there is one time as we say that there is one fountain: the water changes and the fountain remains the same because the form is preserved; the form is preserved because each successive wave takes over the functions of the one that preceded it: the wave that first pushes the wave that it pushes becomes in turn a wave that is pushed relative to another . . . It is here that the metaphor of the river is justified, not inasmuch as the river flows but inasmuch as it makes a unity out of itself.[22]

On the same page Merleau-Ponty offers what is at least verbally a different reason why time remains the same:

> time remains the same because the past is a former future and a recent present, the present [is] an imminent past and a recent future, the future, finally, [is] a present and even a past to come. That is to say, [time remains the same] because each dimension of time is treated or aimed at *as* something other than itself.[23]

To understand all this will be to understand why these various descriptions of why (how) time remains the same are equivalent, though not equally deep. What is involved at any level is a description of the production of identity through repetition; thus, time remains the same because its retentional-impressional-protentional form remains the same by repeating itself. But this form does not remain *statically* the same, as would be the case if it were "a form in the Kantian style, ideally separable from its matter". We may say that the form *becomes* the same, through each 'wave' (phase) repeating the functions of its predecessors. However, this functional repetition has its ontological ground in the modal repetition (designated by *as*) of the present: the future is a future present which is repeated as the past *of* a new present, as what that new present *used to be*, as an aimed-at dimension of itself outside of itself. The form is preserved because the present is. This ontological level of analysis of the time-flow, where the sameness of time is interpreted as the constancy of the present through its changes of mode of being, is not touched by Husserl, who, as Sartre says, "remains timidly on the plane of functional description" (EN, 115): with Husserl the form remains the same because the functions are identically repeated. (I think Merleau-Ponty is wrong to separate the interpenetrative aspect of the time-flow from the dissociative aspect in saying, in effect against Sartre, that the river metaphor images the first but not the second. Retention opens up a *distance* – intends "something other", as he says himself – across which it preserves, as modally changed, a present which threatens to flow *away*.)

It is evident that the type of immobility of the fountain (or river) is not in the least like the 'movement' of a fence across a farm: the fountain moves and does not move, it achieves immobility by *its* motion, whereas the fence across which *our* motion-imparting (tensing) eye allegedly moves does not itself move. Correspondingly, the paradox of the fountain's motional structure is formally irresolvable, whereas the apparent paradox of a 'moving fence' or 'road receding into the distance' is easily explained away as the effect of a verbal shorthand which treats the tensed perspective on a system of tenseless relations as if it were intrinsic to that system. This suffices to give us a glimpse of how the phenomenological conception of the difference between temporal stasis and temporal dynamism differs from the analytical conception of the difference as that between tenseless and tensed time. The phenomenological stasis emerges through dynamism (see above, pp. 49, 57), the tenseless stasis excludes dynamism.

The second objection is that time-motion spatializes time. This presupposes, first, that 'time flows', 'time moves', etc. assimilate time to motion, which they do – for lack of a language specific to time, but second and wrongly, that motion is adequately analysed as the fact of a body being at one place at an earlier time and at another place at a later time. The metaphors do express time in the symbol of motion, but without thereby spatializing time, because the motion to which the river metaphor assimilates time is real motion which is comprised of timelike motions (somehow) and not of the series of positions moved through ("distance *in space*, covered"). So the charge of spatialization rests on an atomistic caricature of real motion as being, in Bergson's phase, "a construction out of immobilities".[24]

The third objection is that any presumed time-motion would take place in an ulterior time-motion, whence a vicious infinite regress of time-motions. So the logically decent thing to do would be to restrict the concept of motion to physical things which move in a time that does not move (pass). But this, like the charge of spatialization, is not the result of postulating a time-motion but rather of assimilating time-motion to atomized physical motion, by assuming that any time-motion would have to be *in* (measurable by) time just as any physical motion must be. Once this assumption is made – mechanically trusting to the univocity of the word 'motion' rather than 'looking at the phenomena' – then it logically and lamely follows that: *just as* any motion of a physical thing must take place in time, so any motion of time would have to take place in time; and so we get the vicious infinite regress that is commonly supposed to scotch any notion of time passing. Whereas Husserl nips the regress in the bud by denying that time-motion is in the least like infinitely divisible motion in space. "Is it not inherently absurd to regard the flow of time as an *objective movement*? *Certainly!*" (ZB, 333; emphasis original).

But what can be said positively of Husserl's time-motion that does not guy it as a species of physical motion? Very little. We cannot *straightforwardly* say that it changes or persists. Change and persistence are for Husserl mutually exclusive modes of endurance of what is in time. Endurance, thus change and persistence, are concepts applying only to transcendent things and processes constituted by time-consciousness, not to time-consciousness itself: "endurance is the form *of an enduring something*", whereas "in the original flow there is no endurance" (ZB, 113; emphasis added). So although time remains the same, this 'remaining the same' is not to be taken in the ordinary sense in which we say that the colour of an object has remained the same. The continuous identity of Husserl's absolute time-consciousness is "not the identity of something persisting" (ZB, 114). As for the changing mode of endurance, although we have to say that time 'changes' (passes, moves), this changing is again not to be taken in any of the familiar senses. Time does not change from being thus-and-so to being otherwise: time, or at least serial time, is homogeneous, despite (or rather because of) its flow. Moreover, ordinary non-temporal change and persistence are interconvertible, in that what has persisted may start to change and what has been changing may start to persist, whereas the sense in which time remains the same excludes the possibility of its starting to change, and the sense in which it changes excludes the possibility of its starting to persist.

Nor can we say that time-motion is a process, since process presupposes persistence. "Objective time", says Husserl, meaning the constituted time of things and events

> is a form of persistent objects, of their changes, and of whatever other processes go on in them. Process is thus a concept which presupposes persistence. But persistence is a unity [of a thing's or event's phases] which is constituted in the flow, to whose essence it belongs that there can be no persisting in it.
>
> (ZB, 113)

If I have so far proceeded by negations in attempting to characterize time-consciousness, it is because the 'change' and 'remaining the same' that apply to it are ultimately ineffable in positive terms. In one mood Husserl resorts to calling time-consciousness "timeless" (ZB, 114), an epithet which misleadingly suggests that time-consciousness in *no* sense changes. In a more frequent mood he has it undergo inverted-commas change:

> Essentially no phase of this flow can be expanded into a continuous succession; and therefore the flow cannot be conceived as so transformed that this phase would be extended in identity with itself. Quite on the contrary, we necessarily find a flow of continuous 'change'.
>
> (ZB, 74)

Husserl's problem is of course that "for all this we have no names" (ZB, 75), since the only words available apply to the time of constituted objects (flowing *in* the river) and not to the self-constituting stream of the time-consciousness (the flow *of* the river) in which they are constituted. Time-consciousness can only be represented "pictorially, as a river/flow" (ZB, 75). The unsympathetic reaction would be to insist (with F.P. Ramsey) that if you can't say it you can't say it and you can't whistle it either. All one can do in the face of this attitude is induce the objector to fix attention on a river or fountain and let *it* speak out its unsayable 'change', taking short breaks to ingest Merleau-Ponty's remark: "[i]n the silence of original consciousness we witness the appearance not only of what words mean but also of what things mean".[25] Who could deny that the fountain changes, yet otherwise than from some state to some other? Who could deny that the fountain remains the same, yet otherwise than in the inert way the pen in my hand remains the same for as long as I look at it?

In particular, time-consciousness does not change in the sense that parts of it would become past and new parts present: "no part of the flow can change into non-flow" (ZB, 114). In that time-consciousness flows, it is not static. In that no part of it is present when another is past, the flow of time is not tensed. The temporality of consciousness and that of its objects exist on different dimensional levels. What can be said to be present, past or future in the ordinary predicative (tensed) as distinct from pre-predicative sense is our constituted acts (immanent objects) and the transcendent objects intended in them. As we saw, for Husserl present past and future are modes of appearance of (phases of) a consciousness which does not itself appear in these modes:

> *the flow of the modes of consciousness is not a process: the consciousness of the now is not itself now.* The retention that exists 'together' with the consciousness of now is not 'now', is *not simultaneous* with the now, and it would make no sense to say that it is. It will be said . . . that retention is something that has its own now, and the same now as a tone, for example. No! There lies the basic mistake.
>
> (ZB, 333; emphasis original)

An event appears to consciousness as now, but the consciousness to which it appears as now does not share that now; if it did, it would appear to itself as an object, as what it is not. But we have seen (above, pp. 48–49) that the fact that it does not appear to itself as an object does not mean that it does not appear to itself at all.

Where does this leave Gale's earlier recorded claim that "events are *continually* becoming present" as an illegitimate expression? I think it should now be evident from what we have learnt of Husserl's theory that the expression need not be taken in the regressive way. To say that events become continually present (future and past) is to say that they become present (etc.) *in the continuity of a time-consciousness which is not a duplicating meta-series of tensed events*. We have, indeed, a time wrapped up in a time, but this leads to no regress because the containing (constituting) time is dimensionally different from the contained time. This consideration will come into its own in Chapter 5, when we consider McTaggart's paradox in the light of Husserl's theory of time-consciousness.

Section 2 – Incurable time

Notoriously, problems about time proliferate the moment we reflect on time-experience, or look for conceptual consistency between our ordinary temporal locutions. But it matters that language and reflection can be cited as alternative sources of this perplexity; for it is a sound principle of philosophical reasoning that answers to problems must, on pain of missing the point, be given in the same context as that in which they arise. If, for example, logico-linguistic analysis showed the notion of tense to be self-contradictory, as McTaggart claims (see Chapter 5), it would be an act of scientific imperialism to lay it down that the fact that a favoured interpretation of time in relativity-physics dispenses with tense means that there is no real problem.

Some clearing of the ground is in order before we can arrive at a sense in which ordinary language and reflection might be named rival culprits in the production of philosophical problems about time. In the first place, it has to be conceded that between them they could never hold the field. For problems also arise within science and in the semantic gap between pre-scientific and scientific conceptions of time. Yet though we might be baffled as to how the 'time' of relativistic physics connects with what we ordinarily understand by the word, this does not put our ordinary understanding of time in contradiction with itself. It puts it out of joint with a non-ordinary understanding. On the other hand, something like self-contradiction, better described as an eclipse of self-understanding, is the stigma of any reflective disclosure of time as a problem. We expect to be able to mean what we say; yet it is a truth universally acknowledged that temporal talk is Pickwickian, in that we never can really mean what we say. When we say that parts of an event are past, we think we mean that they have ceased to be. When we say that we measure an event, we think we mean that we measure the whole of it. And when we say that the dinner lasted an hour, we think we mean that we have measured its duration. Yet we discover on reflection that we cannot consistently mean to say all three: that the past parts of the dinner are non-existent; that to be measured the dinner must exist as a whole; that we have in fact measured the dinner's duration. Now this sort of eclipse of self-understanding is clearly triggered by reflection on meaning-to-say, on intentional speaking, not by reflection on a pure pre-linguistic time-consciousness. Considering that this reflection on intentional speaking involves language, it might seem as if our mooted aetiological disjunction between reflection and language cannot be sustained. And yet it can. For it does not depend on all reflection having to be on a pure pre-linguistic consciousness: reflection on intentional speaking is a perfectly good special case of reflection. The sense in which language is fundamentally opposed to reflection is provided by the wholly non-intentional notion of language-use as a reified quasi-systematic behavioural mechanism whose occasional malfunctioning induces problems not thrown up by reflection and so not experienced in a reflective intuition of 'the light of pre-reflective understanding going out'. This conception of language-use as (mis)behaviour is germane to the Wittgensteinian story as how *pseudo*-problems arise from the unconscious holding of false beliefs about the logic of language games.

I want to argue (A) that in so far as problems about time arise independently of science they are fundamentally connected with reflection, though they find their natural expression in language; and (B) that the Wittgensteinian diagnosis of their origin in ordinary language fails to generate the requisite perplexities. In arguing both these claims I hope to contribute to a defence of the relevance of phenomenology to the aetiology of *genuine* perplexity about time.

(A)

I shall state my case in the major context in which the Wittgensteinian diagnosis has had its say, namely Augustine's famous lament: "What then is time? Provided that no one asks me, I know. But if I want to explain it to an inquirer, I do not know."[26] Reactions to this in the tradition of commentary have been split between those that embrace and those that attempt to dispel the odour of paradox emitted by the clash between "I know" and "I do not know". I shall myself embrace the difficulty, which I shall characterize in terms of the phenomenological theory of reflection. This means starting from the innocuous assumption that Augustine's not knowing what time is is a state he finds himself in as a result of something or somebody having prompted him to reflect.

This characterization of the situation will not issue in a solution or dissolution. But nor will it leave a *problem* hanging. What will emerge is that the difficulty to which Augustine's dictum testifies is *necessary* to human consciousness being the way it is. Calling this unavoidable situation a problem would be like describing as a problem the biochemical necessity that cells must die by living, if not by accident. A problem must be soluble in principle, whereas swapping this world for one in which death is not the price of life is not a solution open to us. For the cellular situation as for Augustine's, the word 'predicament' (rather than 'problem' or 'paradox') seems the most appropriate. Augustine is caught in the human predicament, in so far as this is determined by the nature of consciousness.

What exactly is his predicament? Let us first dispose of the useful but untenable hypothesis that it results from a banal lapse of memory, scarcely deserving of a place in history. It is admittedly true that memories are often prompted by inquiries. 'How is John?' somebody asks, and I remember that he was not so well when I last saw him. But we can imagine the inquiry not triggering that (or any other) memory. Indeed, we can imagine nothing ever reminding anyone of any past event. More than that, we can imagine not having any memories at all, triggered or spontaneous. Unlikely, of course, but shown to be apriori (eidetically) possible by the fact that we often live through stretches of experience containing no rememberings. I know that I have had no memories so far this morning, for example. Yet I have not been unconscious. In short, memory, though a frequent visitor to consciousness, is not necessary to its essence. (It is indeed necessary to consciousness to have a retained past, but we saw earlier (p. 45) how this differs from the remembered past.)

Just because memory is not essential to our self-understanding as necessarily having past, a lapse of memory does not 'blow our being', leave us bereft of something we understand to be necessary to our conscious existence (though the erasure of all retentions would do that). Yet Augustine surely feels bereft of something fundamental. But that this is not his past is confirmed by the fact that he affirms the reality of his past, in claiming *that he knew* what time is. His past is there all right, reflectively affirmed in counter-distinction to his ignorant present. Only he no longer knows *what* he knew. If Augustine knows for sure *that* he knew, then the reproductive act in question is reflection (the certainty of the reflective cogito), not dubitable memory. And his predicament is the conjunction of knowing that he knew and no longer knowing what he knew, in a context of self-understanding which requires that the first be conjoined with the second. The reflective-theoretical framework of interpretation will accordingly have two strands, the one relating to the epistemological surface of reflection (the *that*), the other to the ontological meaning of the reflective bid for appropriation of the *what*. The immediate question arising from the epistemological perspective is the justification of the certainty of the past as accessed by reflection.

Is it possible to motivate radical scepticism about reflection on the model of radical scepticism about material objects? Well, anyone to whom that sort of scepticism makes sense is committed to the view, rejected by the realist, that we do not owe our concept of a material object to any knowledge of the extra-perceptual existence of material objects (whence the *prima facie* plausibility of logical positivism's programme for translating the physical-object language into a sense-data language). Construed by analogy with radical scepticism about material objects, radical scepticism about reflection (for simplicity's sake I disregard lesser scepticisms) rests on the assumptions: (i) that we do not owe our concept of pre-reflective life to any knowledge of it we possess independently of reflection; and (ii) that (i) justifies the wholesale modelling of the relation between reflection and pre-reflection on the relation between perceptual judgments and the extra-perceptual objects they purport to know; the notion that pre-reflective consciousness independently precedes reflection could be an illusion produced by reflective consciousness just as the existence of extra-perceptual objects could, according to the sceptic, be an illusion of perception.

(i) may be conceded to the sceptic, before rounding on him or her over (ii). It is true that before we reflect we have no inkling of pre-reflection (or of reflection). Pre-reflection does not possess the concept of pre-reflection, for it is only reflection on consciousness that judges consciousness, forms concepts generally. As Sartre says of his pre-reflective cogito: "[t]he immediate consciousness I have of perceiving does not permit me to judge" (EN, 19). The concept of a consciousness before reflection can only be gained from reflection.

It turns out, however, the concept of pre-reflection gained from reflection forbids extending the perceptual analogy to the point where it would give a foothold to scepticism. For the testimony of reflection is that 'pre-reflection independently precedes reflection' means, not that pre-reflection independently precedes reflection *independently of reflection*, but that pre-reflection independently precedes reflection *for reflection*. That tension 'independently of reflection for reflection' is attested in the ambiguous mode of givenness of pre-reflective consciousness to reflecting consciousness: *as reflected*, the pre-reflective consciousness gives itself as sustained in existence by the reflecting consciousness; while *as pre-reflective* it gives itself to reflecting consciousness as having independently (and namelessly) preceded reflecting consciousness, but as having lost its autonomy (EN, 199).

Now it is clear that 'material objects exist independently of perceptual judgments only for perceptual judgments' is not what perceptual realism claims. And it is equally clear that, conversely, 'pre-reflective consciousness independently precedes reflecting consciousness independently of reflecting consciousness' is not what reflecting consciousness claims. In other words, the perceptual analogy is deeply defective here because the claims of reflection and of perceptual judgment are not structurally homologous.

The lack of homology is due to the fact that in the one case we are dealing with a relation of consciousness (as judgment) to putative exemplars of non-self and in the other with a relation of consciousness to *itself*. To say that pre-reflective consciousness is *for* reflecting consciousness is to say that it gives itself as what the reflecting consciousness ecstatically *was*, or that the reflecting consciousness is what the pre-reflective consciousness *has become*. (That these modal descriptions should apply again here attests to reflecting consciousness being itself pre-reflective, therefore subject to the laws of primary consciousness.) In short, the sceptical criticism does not preserve the essence of reflection while undermining its claim to certainty: it obliterates the essence of reflection. A consciousness that satisfied the condition on certainty it imposes (absolute precedence) would simply not be a *reflecting* consciousness, because it would intend an *object*, a target it might miss.

Because the reflecting consciousness *was* the reflected consciousness, Husserl can say that "the reflecting act includes its object within itself to such an extent that it is only by abstraction, as a moment incapable of standing by itself, that it can be separated from it".[27] To such an extent, indeed, that strictly speaking reflection does not posit consciousness as an *object* of knowledge, thus as something presenting itself through perspectives and about which we could be mistaken. As Sartre puts it: "reflection is a recognition rather than a cognition" (EN, 202); an intuition invaded and blinded by its "quasi-object" (EN, 119), which presents itself all-of-a-go and without perspectives.

In contrast to the case with reflection, the objects of memory are not 'included' in acts of remembering. We graciously accept the fallibility of memory because we know that the act of remembering can survive the non-existence of the object intended in the act. I cannot doubt the reflective judgment that I have been feeling a headache, though I can doubt my memory of having taken an aspirin for it this morning (doubt what is remembered, that is, not the existence of the reflectively verifiable act of remembering). Memory and reflection are species of "inner perception" which Husserl defines as "those acts to whose essence it belongs that [their] intentional object, *should it exist*, belongs to the same stream of consciousness as the act itself".[28] We know from eidetic intuition (not from factual checks) that memory *can* fail to satisfy this existence-condition, whereas we have seen that reflection satisfies it as a necessity of its essence. Factual checks make sense only in the context of the admitted logical possibility of error, which is why the notion of doubting or checking up on my reflective judgment that I have been feeling a headache is senseless.

So much for understanding in terms of phenomenological theory of reflection how Augustine can know for sure *that* he knew what time is. Understanding how this is compatible with his failure to know reflectively *what he knew* requires that we get clear about what we expect from reflection, the standard by which it is a partial failure. The following over-condensed remarks from Merleau-Ponty will take us some of the way.

> The movement of return to ourselves – of 'going back into ourselves', as St. Augustine said – is, so to speak, rent by a reverse movement *which it gives rise to*. Husserl rediscovers this identity of 'going back into onself' and 'going out of onself' which, for Hegel, defined the absolute. To reflect, as he said in *Ideas I* is to disclose a pre-reflective consciousness which is at a distance, since we are no longer naively at one with it, and which we nevertheless know without doubt to be reached by reflection, since it is through reflection itself that we have an inkling of it. It is therefore not pre-reflective consciousness itself that thwarts [*conteste*] reflection, it is reflection that thwarts itself because its attempt at recovery, at possession, interiorization or immanence, has by definition meaning only in relation to a term which is already given but which withdraws into its transcendence beneath the very [reflective] gaze which chases after it.[29]

The emphasis here is on the idea that the identity between the reflecting and reflected consciousness is ecstatic, identity *at a distance.* The purpose of "going out of onself" is to return to oneself from a distance, all the better to appropriate oneself. Augustine's testimony, however, implies that this aspiration is somehow thwarted. And we have just noted Merleau-Ponty say that the fault does not lies with the pre-reflective consciousness – it is not as if reflection were prevented from taking its distance by the refusal of pre-reflective consciousness to unglue itself from itself. The suggestion is rather that there is something about the reflective distancing that does not go quite according to plan: there is a *reculer*, but it does not pull off its *mieux sauter*. What we want to know is, Why not?

We shall see that the reason is logically connected with the fact that the return establishes reflective certainty – the direct contact of consciousness as reflective with consciousness as pre-reflective. As we have seen, it is this indubitable cogito which enables Augustine to know for sure *that* he knew what time is. But his testimony suggests that this return is not also such as to *present* the pre-reflective consciousness known with certainty to have been 'knowledgeable about time'. Supposing that presence were achieved, however, then not only would reflection achieve an identification across time of consciousness with itself – an identification between earlier and later phases of the same time-stream – it would also achieve an objectivation of consciousness by itself, such as would enable Augustine to know propositionally what time is. We would have a unity of self-interiorization and self-objectivation, of certainly and thematizability (propositional knowledge). This is absolute mind, the twofold standard to which reflection aspires. Here is Sartre's statement of the standard:

> the motivation of reflection consists in a double and simultaneous attempt at self-objectivation and self-interiorization. To be for oneself an object in itself within the absolute unity of interiorization – that is what being-reflection requires to be.
>
> (EN, 200)

Sartre has shown how reflection succeeds in respect of certainty while failing in its bid for self-objectivation (see his brilliant discussion of "the certainty of reflection", EN, 201–205). The aspiration of reflection, as of all human action according to Sartre, is to be like a stone which *per impossible* manages to be *present to itself* in its infinitely dense stonehood: to achieve a contradictory fusion of temporal being-for-itself with timeless being-in-itself. In that consciousness still desires to be consciousness, it does not desire timelessness, but rather to have mastered and appropriated its time, to have "saved [it] from ecstatic dispersion" (EN, 200), whence the popularity of 'full life cover'. The idea is that we "*be* that flight instead of temporalizing it as the flight that flees itself" (EN, 200). Yet "this effort to . . . recover and dominate within itself its own flight must end in failure; and it is precisely this failure which is reflection" (EN, 200). The necessity of this failure is ordained by the contradictoriness of the mentioned ideal fusion. But we need to grasp that contradiction in terms tailored to our purpose.

The reflecting-recovering consciousness is not a thing but a for-itself temporal consciousness. Since it desires to recover *itself* in recovering pre-reflective consciousness, the pre-reflective consciousness must be recovered as also being a for-itself temporal consciousness; for otherwise the recovering consciousness will not acknowledge identity with the recovered consciousness: "the being which operates the recovery must constitute itself in the mode of for-itself and the being which is to be recovered must exist as for-itself. And these two beings must be *the same being*" (EN, 201). Therefore, the condition of self-interiorization (of certainty) is that the pre-reflective consciousness *not* be recovered as a thematizable in-itself object. And in fact it isn't: "pure reflection is never more than a quasi knowledge" (EN, 209). On the other hand, the condition of self-objectivation (of Augustine being able to say *what* he knew) is precisely that the pre-reflective consciousness be recovered as an in-itself *object* present to thematizing reflective inspection. So the satisfaction of the condition of self-interiorization entails the non-satisfaction of the condition of self-objectivation, and vice versa. In other words, Augustine's certain reflective knowledge that he pre-reflectively knew what time is entails that he reflectively cannot know what he knew.

We tend to imagine that reflection succeeds in rendering consciousness present as an object because we suppose (as Descartes did) that it takes us out of time. Do we not 'gather ourselves together' in reflection, arrest the temporal dispersion? Only in intention, as Merleau-Ponty reminds us in connection with the problem of phenomenological reflection:

> if we were absolute mind the [phenomenological] reduction would not be problematical. But since, on the contrary, even our acts of reflection take place within the temporal flow which they seek to capture (since they *sich einströmen*, as Husserl says) there is no thought which embraces the whole of thought.[30]

The reason why the reflective project does not go according to plan is that reflection turns out not to be a perfectly circular movement of consciousness curling up in itself and escaping the temporal dispersion, but rather a movement of return which, simply because it is self-temporalizing like any other consciousness, undoes itself into a movement of departure ("thwarts itself"). If we may dredge another metaphor from the river: a wave attempts to turn back on itself, but is prevented from doing so by the fact that it is made out of the same mass of water as the forward-rushing tide. While all acts of consciousness 'go along with the stream' (are in time), it is the specific thetic character of reflection to go along with it recalcitrantly, to attempt to envelop the time by which it is enveloped. It cannot succeed, because "the one who is reflecting on myself is not some sort of non-temporal gaze, but myself, myself who am enduring" (EN, 199). (It should be borne in mind that Sartre is referring here to the reflection he dubs "pure", as distinct from "impure reflection": pure reflection produces no objects, whereas impure reflection creates the psyche-object together with the psychic temporality we saw Sartre identify with Bergson's *durée*, above, pp. 34–37).

Augustine's self-eclipsing cogito has turned out to be more modern, not to say postmodern, than that of Descartes and the tradition he inaugurated. For the cogito of the Cartesian tradition (of what Derrida calls 'Western metaphysics') expresses, not just an indubitable truth, but an indubitable truth attained in the act of consciousness possessing itself as an instantaneous *presence*. Whereas with Augustine we have certainty ('I knew') without presence ('I know not what I knew'). The very act by which I come to know reflectively that I pre-reflectively knew what time is also occludes the presence, and so the decipherability, of what was known. The reflective light darkens, raising problems for phenomenology's own reflective method. (The good news is that phenomenology has proved itself to have resources for identifying and analysing 'originally absentive phenomena' – Levinas's time, for example – despite its initial methodological assumption that reflection gives access to presence.) My postulate that Augustine's dictum records a shock to his being implies that, although 'I know' and 'I do not know' are merely contradictory statements, the underlying intuitions are better (but still imperfectly) represented by the *self*-contradictory statement 'I know and I do not know', uttered in one breath. That is, the shock implies that the knowing and the not-knowing are not separated from each other as an earlier from a later tensed time. Suppose Augustine never reflected on time before he went to Rome. Then had he been asked in Rome, 'When did you know what time is?' he might have answered either 'Before I came to Rome' or 'Before I reflected'. If the second 'before' is put on a par with the first, the knowing and the not-knowing are construed as states existing at different moments of tensed (public) time, and it becomes unintelligible how Augustine's reflection could generate anything resembling a self-contradiction. For there is nothing self-contradictory about knowing what time is on Monday and not knowing what it is on Tuesday, whereas there is indeed something comparable to self-contradiction about a pre-reflective knowing and a reflective not-knowing being 'extensionally together' (in Husserl's sense) in a time-consciousness of which we cannot say, as we have seen we cannot, that the knowing phase is out to play when the not-knowing phase is present. Self-contradiction, however, is still an inadequate characterization of Augustine's predicament, since it involves the conjunction of mutually incompatible facts *at a tensed time*, whereas there are no tensed times in time-consciousness. Whence my recourse to the metaphor of an eclipse of self-understanding, which cannot be formalized into self-contradiction.

(B)

I have argued that the root of Augustine's perplexity is the failure of reflection, the thwarting of the aspiration of consciousness to achieve the self-presence of absolute mind. With this I go against the traditional diagnosis, according to which Augustine knows what time is as long as he is not called upon to give it a verbal definition. Were this his predicament, there would be force in the oft-made deflationary remark that since plenty of other concepts are verbally (as well as ostensively) undefinable, there is no good reason for special fussing over time. But there is something special about time. We would not be perplexed following a request to describe the ineffable taste of blackberries, though here again we should have to say 'I do not know'. This avowal of ignorance means no more than that we are unable to capture the taste in a definition, despite our 'knowing' it in the tacit sense that we are able to identify it, are never fooled by blackberry-flavoured taste-alikes.

Consider Augustine's dictum with the appropriate substitutions. 'What then is the taste of blackberries? If no one asks me I can identify it; but if I wish to describe it to an inquirer I cannot do so by means of a verbal definition.' The odour of paradox has evaporated. There is nothing paradoxically perplexing, if perplexing at all, in the fact that much of what we experience as distinctive cannot be put into words. The two cases have in common, however, that the reflecting consciousness 'occurs' within the same stream of time-consciousness as the pre-reflective experience it consults. But there is this crucial difference: that although reflection on the taste of blackberries remains immersed in the same stream of time-consciousness as the pre-reflective tasting consciousness, it does not attempt to thematize that pre-reflective consciousness in respect of its temporal structure – it thematizes it *against the unthematized horizon of its temporal structure.* As such, it is a less radical exercise of reflection than Augustine's, which attempts to thematize the temporal ground on which it stands. Our inability to define the taste of blackberries derives wholly from the limitations of language, whereas the reflection that attempts to appropriate time-consciousness as such finds that this has *absconded*, not that it is adequately present to mind as something for which we cannot find words. Time is not 'there' to be caught and defined when we peer after it.

A more recent linguistic approach deriving from Wittgenstein traces the perplexity to the demand for a definition of time falling foul of the fact that time is not the sort of thing for which it is logically proper to make such a demand. On this showing, it is no longer deemed a pity that time cannot be defined, but wrong-headed to expect that it should be. It is argued that when Augustine asks 'What then is time?' he is under the twofold illusion that all nouns, including 'time', are names and that all names get their meaning from the things they name: not just 'Fido' from Fido. His perplexity is thereby diagnosed as arising from his misguided search for the "queer process" (Wittgenstein) he supposes must exist in order to make the name 'time' mean something. To this it may be replied that, even granting that both these semantic beliefs are false and that Augustine held them, it still would not follow that he would have been wrong to suppose that time does not possess some sort of being. All that would follow is that he would have had bad reasons for believing that it does. And that "a queer process" is not the only option open to someone who believes that time is 'not nothing' is attested by the fact that every phenomenologist save Bergson explicitly denies that fundamental time is a process (see above, p. 126; also below, pp. 192–193, Heidegger's critique of the notion of time as "something occurrent").

More fundamentally, however, it may be argued that perplexity over time could not arise from holding the sort of beliefs, true or false, conscious or unconscious, that hold sway in analytical epistemology. The difference between these analytical beliefs and the intuitions of phenomenology may be characterized in terms of the different concept of evidence involved in either case. The evidence for an analytical belief is its justification, which in the case of perceptual beliefs is either whatever fundamental items are in favour (sense-data, basic propositions, etc.), or else other beliefs cohering within the same belief set (or both). The important point is that the evidence is always external to the subject-matter. I infer a nearby dog from a "canoid patch" (Russell), the dog is not immediately evidenced with its coat, as its coat. Suppose, now, that 'I know what time is' and 'I do not know what time is' are taken as recording beliefs of this type. Then we might try to generate perplexity about time by saying that the contradiction between the two beliefs justifies an inference to the further belief that time is perplexing. It seems to me, however, that there is no good reason for preferring this inference over the inference to 'I am a fickle (forgetful) believer', unless it has been admitted that time(-consciousness) has *evidenced itself* in contradictory style to reflective intuition.

The Wittgensteinian dissolution of Augustine's predicament admittedly does not interpret 'I know' as a belief but rather as a pre-theoretical competence with the logical grammar of temporal expressions. The negative claim here can hardly be disputed, in that it is true enough that Augustine's pre-reflective 'knowing what time is' is not a state of propositional knowledge. Phenomenology, however, accommodates Wittgenstein's disclaimer with its notion of pre-propositional (pre-predicative) experience, which has the advantage over 'competence' of not throwing the conscious baby out with the propositional bath-water. The result of Wittgenstein's doing so, in making 'I know' refer to an unconscious behavioural competence, is the loss of the temporally continuous (*same*) consciousness in the absence of which the perplexity for which he offers a diagnosis and a remedy (therapeutic "reminders") cannot even arise. For as we have seen, the not-knowing due to the reflective non-presentedness of time is perplexing only because it arises 'together' with an indubitable knowing that the same consciousness did know, in 'its' pre-reflective past. *Either* we hold on to the postulate that 'I know' refers to a purely behavioural competence, in which case a consciousness reflecting on the malfunctioning of that competence will no more find itself in conflict with it than my incapacity to explain how I produce grammatically correct utterances finds itself in conflict with my knowledge that I do produce them. The fact that I do not know, in the sense of being able to formulate, the rules exemplified by the correct utterances I produce might well perplex the desire to understand just what tacit linguistic knowledge is. But as long as that knowledge is construed as some sort of purely behavioural competence (like riding a bike), any perplexity here will be of a very different kind from Augustinian perplexity: for the move from 'just doing it' to 'not knowing how I do it' involves no eclipse of self-understanding of the sort that characterizes the conjunction of 'I know' and 'I do not know' in Augustine's predicament. So on the behavioural postulate we have no 'cramp' in need of linguistic therapy. *Or else* we take 'I know' to refer to a lost pre-reflective knowledge of time, in which case we have the perplexing clash of this 'knowing' consciousness with the ignorance of the *same* consciousness as reflecting. But then perplexity arises through reflection, independently of linguistic malfunctioning. So in the first case no relevant perplexity arises, while in the second it arises in a way Wittgensteinian therapy cannot address. In neither case, therefore, is therapy indicated.

5 McTaggart and Husserl

Introduction

"The real is not real; or that which is is that which is not." With this pronouncement on what change would mean, Parmenides inaugurated a philosophical tradition which in recent times has had its most celebrated exponent in the Cambridge metaphysician J.E. McTaggart. Based as it is on innocuous premisses, superior in rigour to its predecessors, McTaggart's disproof of the reality of change, and so of time, has supplied the conceptual framework and some of the vocabulary of most of the contemporary Anglo-American debates on time that do not take their cues directly from science. As with all paradoxical arguments, almost nobody believes at heart in the truth of McTaggart's conclusion, though some commentators have defended the validity of his argument. Most have claimed to refute it.

It might seem, even in advance of considering McTaggart's argument, that he is already refuted by the concept of eventless time. If it is accepted that there really is an eventless time in which "no thing changes", then does it not follow that the contradictoriness of change is no impediment to the reality of time? This response would chicken out miserably from engaging with the event-bound Aristotelian time which McTaggart and his commentators have in mind in their discussions of change. A much bolder and more interesting response would be along the lines that event-bound time can be real *because* eventless time is real. This would give phenomenology of time a chance to prove its mettle in the heartland of analytical philosophy of time.

A phenomenological refutation of McTaggart would be all the more significant if it could be made against a background of agreement that he has not been answered by any of his analytical critics (the seminal phenomenologists give no sign of having heard of him). I believe he hasn't, though it would take another book to prove the point by the method of climbing over the mountain of commentary on McTaggart. I shall therefore take it as read that only two conceptions of change (the tensed and tenseless) are on offer in mainstream analytical philosophy of time, and show by reference to a few proffered 'answers', that any answer based on these conceptions must suffer from a flaw congenital to both, namely, the substitution of pseudo-succession for change. A qualified phenomenological solution will be attempted in the final section of this chapter. Meanwhile, before rescuing time from McTaggart, I propose to spend some time rescuing McTaggart from his analytical critics – fattening him up for the kill, as it were.

McTaggart's argument[1]

The temporal positions of events appear *prima facie* to come under two descriptions. First, any event is earlier than one set of events and later than another set. So considered, events are ordered in a tenseless series dubbed "the B-series" by McTaggart. Second, any event has a temporal position in relation to the present: it is present or past or future. So considered, events are ordered in a tensed series dubbed by McTaggart "the A-series". When I say that the First World War is earlier than the Second World War I refer to these events as members of the B-series, whereas when I say one is present, or that both are past (etc.), I refer to them as members of the A-series. Now, change must be a characteristic of at least one of these series if time is to be real, for it is one of McTaggart's innocuous premisses that change is essential to time. No change is located in the B-series since the relations comprising it never change: if the First World War is ever earlier that the Second World War it is always earlier than it. The fact that the B-series has an order no more implies change than the order of the letters of the alphabet does. So "the B-series by itself is not sufficient for time, since time involves change" (M, 461). From this McTaggart does not immediately conclude that the B-series is unreal. What he concludes is that it could not be real off its own bat. The tenseless B-series could be real only by virtue of parasitic correlation with an A-series whose terms would be events undergoing change from future to present to past.

So whether time is real or not hangs on whether the A-series qualifies for reality by accommodating change. Now, our concept of tensed change demands both that every event be present, past and future and that these three "A-determinations" (to borrow Gale's expression) be mutually compatible. But in fact they are mutually incompatible. Our concept of tensed change is therefore self-contradictory. What is self-contradictory cannot be true of reality (another premiss). So no A-series can really exist; and since the B-series could only exist by courtesy of it, it follows that no B-series exists either. Time is unreal – a misperception of an ordered but non-temporal C-series belonging to the higher reaches of McTaggart's idealist metaphysics.

McTaggart anticipates the obvious objection that, while change requires that an event be present, past and future, it does not require that the event possess these incompatible A-determinations simultaneously. Change is possible, one might suppose, because events possess the A-determinations successively. Then we are saying that event M is present-and-only-present (present *simpliciter*) at a moment of meta-time which is present. But with this move we have simply transferred the original contradiction from an event to a moment which is, again, past and future as well as present. Since the same problem recurs with every move to a higher meta-level, it is clear that we are embarked on a vicious infinite regress, symptomatic of the false premiss that there can be a time when an event is present *simpliciter*. There can be no such time, there can be no time when the truth conditions of 'M is present' obtain and those of 'M is future' and 'M is past' do not. There can therefore be no unique designation of the present such that M bears a changing relation to it, whence it follows that there can be no tensed time (no A-series).

McTaggart's series of earlier/later relations is the series I have called "becomingless" (see above, p. 112) – tenseless as becomingless. No event enters this time-series by dint of coming to pass: "N will always have a position in a time-series and always has had one. That is, it always has been an event and always will be one, and *cannot begin or cease* to be an event" (NE, 12; emphasis added). With this series we have no difficulty assigning events to time-positions (assuming the series to be genuinely temporal), we simply say that event M is later than L and earlier than N and answer the question, 'When is that true?' with 'at all B-times'. But does this not suggest that the problem with the A-series may be solved by answering the question, 'When is M present and only present?' in terms of the tenseless relations generated by the A-series. Might the question not be answered within the first-order A-series by remarking that it is a tenseless truth that M is present when L is just past and N just future? No, for that tenseless truth (as distinct from the tenseless truths applicable to the B-series construed as becomingless) *starts* to hold as from when M is present. But that means that it starts to hold forevermore only once M *becomes* present. But the time when M becomes present and the mentioned succession starts to hold forever after clearly cannot be fixed in terms of the earlier/later relations constitutive of the succession itself. Yet any moment we may care to nominate as one when the succession holds will be subject to the unanswerable question, 'When is that moment present and only present?' Successions are no better than events at having a unique position (extension) in time.

McTaggart's concept of change

Might it not have been possible to save the reality of change by conceiving it otherwise than as the change of A-determinations of events? By what right does McTaggart tie change to just that sort of change? Well, suppose we have some M which 'changes', as we say *into* N, as Hyde into Jekyll; or which is simply *replaced* by N bearing no trace of M: are we dealing here with genuine species of change?[2] Not if change requires that whatever changes must come out of the change identically the same as it went into it: to be able to say that an apple has changed from green to red, we need to be able to point to the red apple and say, 'That is identically the same apple as was green.' McTaggart affirms this ancient common-sense principle of the preservation of identity through change apropos of events: "changes must happen to events of such a nature that the occurrence of these changes does not hinder events from being events, and the same events before and after the change" (M, 260). That 'hindrance' I take to allude to the notion that a putative change which obliterated its subject, replacing it by some second entity, would be no change at all.

Taken as its stands, however, McTaggart's affirmation of the principle of preservation might seem to leave out the fact that, while no thing can change unless it remains the same, neither can it change unless it comes out of the change *different* from the way it was going into it. A change that made no difference to it would no more be a change than a change which made a difference so great as to obliterate its identity. But now we must take seriously the fact that McTaggart is talking about events, as distinct from things. An event is the making of a qualitative or quantitative difference to a thing (or person). The death of Queen Anne (McTaggart's example) is an event making a difference to Queen Anne. Is any difference made to that event by its changing its position in the A-series? Could any contemporary witness of the death have intelligibly said, 'This death, now that it is present, has different causes and effects from what it had when it was future'? No. (This does something to justify the concept of a future event, despite its non-existence. If the concept were bogus, nothing we might say could violate it, for it would have no good shape. But when we say that Queen Anne's death is different when present from when future, we know that the absurdity lies not in concept of a future death but in the way we are twisting it.) Or would any historian say, 'Queen Anne's death, now that it is past, is intrinsically different from what it was when present, just because it is past; so for a reason that has nothing to do with historical relativism, we shall never know what it was really like'? No. The point is that the change of A-determinations does not stand to events in the same way as the change of qualities (etc.) stands to things: an event comes out of a change from present to past (the change of time) as strictly and simply the same as it went into it, whereas a thing comes out of a change from one quality (etc.) to another as the same *and* different.

Whether things (as well as events) undergo the change of time is controversial (is a green apple past? Or is it the fact, or event, of its being green that is past?). But if things do undergo the change of time, then we must say: in changing in time from green to red an apple both stays the same and undergoes qualitative difference, while in concomitantly undergoing the change of time from a now-green to a was-green apple, the green apple remains strictly and simply the same in all points of applehood. Of course, there is a *sui generis* 'raw-feel' difference between a present and a past event (or thing, perhaps) – but this is not a difference made by these A-determinations imparting different *qualities* to the event: "if these characteristics [A-determinations] are qualities, then the event, we must admit, would not always be the same, since an event whose qualities alter is not, of course, completely the same" (M, 461). But events do stay always the same though their change from future to present to past, whence it follows that these A-determinations are not names of qualities.

It is true that McTaggart overstates his case in identifying change with just the tensed change of events:

> what characteristics of an event are there which can change and yet leave the event [strictly] the same? There is only one class of such characteristics, namely the determination of the event in question by the terms of the A-series . . . we seem forced to the conclusion that all change is only a change of characteristics imparted to events by their presence in the A-series. (M, 460–461)

"Only"? Surely the change in the qualities of things is change too? The extravagance may be remedied at the negligible cost of moderating McTaggart's thesis from 'all change is the change of A-determinations' to 'the change of A-determinations is the basic form of change'. To do this we merely point out on his behalf that a thing's undergoing a change from one quality to another presupposes, and is defined in terms of, the change from present to past, but not vice versa. To be able to change from green to red, an apple must be able to *have been* green, whereas to be able to have been green it need not change to red or to any other post-green state: it has been green when it has been eaten or otherwise obliterated, just a much as when it has changed from green to red.

I have tried in this section to explain McTaggart's concept of change and show that it is justified, or justifiable. In particular I have wanted to show that it does not involve him treating present past and future as qualities. The reason why I have stressed the latter point more than he does will emerge in a moment.

Objections from tensed theory

One strain of McTaggart criticism agrees with him that tensed change is self-contradictory, while denying this consigns time to unreality. I shall come to this later. Meanwhile I want to defend McTaggart against critics who uphold the non-contradictoriness of tensed change, and accuse him of manufacturing his contradiction though an illicit handling of tense. All three of the basic units of his way of talking about tense have been censured: the words 'event', 'is' and 'past' in e.g. 'event M is past'.

Against McTaggart's events

An undeservedly influential line of criticism alleges that McTaggart's argument incorporates a false and baffling notion of what tensed change is, namely change in the A-determinations of *events*. McTaggart is accused of the category-mistake of appointing events rather than things as the subjects of tensed change. Thus one writer:

> Events, to my mind, are not in that class of entity which can undergo changes . . . although it is correct to say that an event is past it does not follow that anything has actually happened to the event itself, or that the event has changed in any way . . . [t]hings change: when something changes, an event occurs. An event is constituted by something undergoing a change with respect to the state it is possible for that thing to be in . . . Events don't change, things do.[3]

If by 'nothing actually happens to the event itself' is meant that the event does not undergo *qualitative* change, then this objection takes the words out of McTaggart's mouth. If, on the other hand, the objection is serious about no change of *any* kind having occurred, then surely the objection fails. An event that was present yesterday and is past today is an event that has changed. It's no use saying that events don't change for the reason that they *are* changes, happenings to things. It is true that a traffic light changes from red to green and that the event of its doing so does not change from red to green. But that is not because events are not the sort of entity that can change – it is because they are not the sort of entity that undergo *qualitative* change, as McTaggart rightly points out. Ultimately the objection rests on a tacitly assumed universal model of change as the picturesque sensuous change of the qualitative attributes of things. This is what leads to the oddness of allowing that an event can be past ("it is correct to say that an event is past") while denying that this is a fact of change in the event.

Against McTaggart's predicates

I turn now to a corollary objection according to which McTaggart artificially generates his contradiction by treating the allegedly pseudo-predicates 'present', 'past' and 'future' as if they were genuine predicates. It is admittedly difficult to say just how past, present and future exclude each other, given that it is not by qualitative content (as I hinted a moment ago in calling the difference between them *sui generis*). 'Dark' and 'light' exclude each other, as do 'wet' and 'dry', but they do so on the basis of incompatible qualitative content. Perhaps we might say that A-determinations have discriminatory content, that is, content sufficient for them to exclude each other, but no transitive content, that is, such as changes the intrinsic nature of what changes from present to past (etc.). However the point be handled verbally, this fact of contrast remains: wet and dry are contents which exclude each other in such a manner that, if something changes from wet to dry, a difference is made to the qualitative composition of that something; whereas present and past have contents such that, if something changes from present to past, no difference is made to the qualitative composition of that something.

Why insist on this? Because the objection claims that McTaggart was misled into manufacturing his contradiction by the grammatical analogy between 'This is wet' and 'This is past'. The first move is to lay it down, in the face of textual evidence to the contrary, that "[h]e construed the words 'past', 'present' and 'future . . . as if they were quality words with a logic on all fours with that 'of dark' or 'red', or 'crinkled' ".[4] From this bad start the argument continues as follows. Just as 'This is dark' is taken to mean that something that is dark exists somewhere, so 'This is past' is mistaken by McTaggart to mean that something having the quality of pastness must exist somewhere; and since the same something must, by the same reasoning, also have the property of presentness and futurity, it must itself be a timeless substratum for timelessly conjoined properties. And the way to avoid this allegedly factitious problem is to disabuse ourselves of the notion that A-determinations are genuine (logical as distinct from merely grammatical) predicates. We shall see shortly that McTaggart understands A-determinations to be *relations* between events and a timeless *x*. The point meanwhile is just that he could not have been (mis)led to the view that they hold timelessly by (mis)taking 'present', 'past' and 'future' to be quality-words, for he explicitly says that A-determinations are not qualities.

Against McTaggart's verbs

A prolific strain of objection against McTaggart has accused him of illicitly detensing verbs so as to bring mutually incompatible A-determinations into contradictory co-existence. In responding to this objection we must distinguish the factual question: (i) to what extent McTaggart uses verb forms which are detensed in the sense that tensed forms would have been more idiomatic, from the polemical question (ii) whether his argument in any way relies on such detensing as he goes in for.

Of the death of Queen Anne, McTaggart does say: "[t]hat it is a death, that it is the death of Anne Stuart, that it has such causes and such effects – every characteristic of this sort never changes" (M, 460). This has been cited by one commentator in evidence of McTaggart's "obsessional preference for timeless verbs".[5] Yet it is wrong to flog the point that he should have written *had* rather than has,[6] for he is considering the death as a event in the tenseless B-series, from which perspective *has* in the tenseless (as distinct from present-tensed) sense is the logically appropriate verb form. The idiomatic forms recommended by the objection would have been conceptually inappropriate, since to say that Queen Anne's death *had* such causes is to suggest that the tenseless truth 'Queen Anne dies from c' has ceased to hold by the later time of writing (frustrating historians). McTaggart has eliminated tense here, but by way of correcting ordinary language in the interests of logical perspicuity, which is served by his *has* in just the same way as it is served by saying, unidiomatically, that the death of Queen Anne *is* earlier by 187 years than the death of Queen Victoria. So if we have here a case of detensing, it is relative to a logically defective convention of English idiom, not to any standard of philosophical correctness. It is therefore not a case of *illicit* detensing.

Anyone can verify for himself that the grammatical facts of McTaggart's text do not bear out the hit-and-run claims that he always detenses. He does and he doesn't,[7] he has no "obsessional preference". And it is hard to see what he could have gained by consistent detensing anyway. The way a philosopher expresses the necessity of every event or moment being present, past and future will tend to exploit whatever resources the language makes available for recording his or her *metaphysical decision* as to whether that necessity entails a contradiction. English, for example, though not some other languages, offers an option between a tenseless formulation suggesting a contradiction and tensed formulations suggesting none. But the matter of substance is the decision, not the mode of expression. The irrelevancy of this whole fuss over tensed vs tenseless formulations may be brought into focus by showing that it is impossible that any way of interpreting 'a moment which is future (etc.)' could be the result of a reading of the 'is'. In that expression all the temporal information is supplied by the word 'future', no independent contribution being made by 'is'. The 'is' is a bare copula, grammatically required by English despite its redundancy, though not by many other languages: 'future moment' is as informative as 'moment which is future'. So any interpretation of 'moment which is future', for example, that it means 'moment which is future and past and present too', takes no account of the 'is' – it is strictly an interpretation of 'future moment'. If there is a contradiction implicit in 'moment which is future', it is already contained in 'moment future', therefore not generated by any (mis)reading of the 'is'. Any incorrect reading could not arise from an incorrect reading of 'is', but only from a wrong metaphysical decision as to how futurity relates to a moment.

It will have been noted from the last quote from McTaggart that the moment to which an A-determination of an event is relativized remains absolutely A-characterized: in 'M is present at a moment which is present', the second 'present' remains unrelativized to a time (absolute, extratemporal). An event cannot possess an A-determination in a relativized way unless a moment can possess it absolutely; which means that the point of any relativization of this kind is defeated by the moment necessarily remaining absolutely A-characterized. Not only is it beside the point whether the relativization is carried out in grammatically tensed or detensed style, the whole undertaking of relativization is pointless. The issue is whether a contradiction is involved in the idea of an event possessing absolutely presentness, pastness and futurity: whether 'M is present' *tout court* implies a contradiction.

One way of generating a contradiction is to treat M as a timeless event-substance which, just because it cannot be at different times, would be inconsistently qualified if it were all of past, present and future. Now, McTaggart certainly does have a timeless way of thinking about time, in the sense that timelessness is a conceptual factor in his argument, but it does not consist in his arbitrarily postulating events as timeless substances. This is more apparent in the version of the argument given in NE, where A-determinations are overtly treated as fundamentally relations, than in the earlier *Mind* version, where they are not positively characterized as to ontological category. So it will be worth restating his argument in the terms of the NE version with a view to pinning down the timeless factor and its role.

It is not a sufficient condition of a series S being an A-series that in it a term L be past relative to a present term M. A series is indeed defined as an *A* time-series by that specific characteristic, but it is not that specific characteristic that defines it, more fundamentally, as a *time*-series. For a time-series must somehow involve change, whereas 'L is past relative to present M' never changes: it remains true when N is present relative to past M, when O is present relative to past N, and so on. No event in the series can change its relation to any other in the series.

To what, then, could an event stand in a "changing relation" (NE, 19), given that the relatum cannot be an event in the A-series? Only to something outside the series, as McTaggart confirms in a somewhat neglected part of his text:

> if, then, anything is to be rightly called past, present, or future, it must be because it is in relation to something else. And this something else in which it is in relation must be something outside the time series. For the relations of the A-series are changing relations, and no relations which are exclusively between members of the time-series can ever change . . . a series is an A-series when each of its terms has, to *an entity X outside the series*, one and only one of the three indefinable relations, presentness, pastness and futurity . . . *This term could not itself be in time* . . . To find such a term would not be easy, and yet such a term must be found, if the A series is to be real.
>
> (NE, 19–20; emphasis added)

So an A-series would be generated by events standing to a timeless entity X in the relation of futurity, changing to the relation of presentness, changing to the relation of pastness. The invariant relations of succession within the series would presuppose these changing relations to something outside the series: there can be no succession which does not rest on change.

"But there is a more positive difficulty in the way of the reality of the A-series" (NE, 20) than finding a timeless X. To any timeless X any event both must and cannot stand in all three mutually incompatible A-relations. It's no good saying that at a given time each stand to the timeless X in just one of these A-relations, since this either triggers the regress or presupposes time in the very effort to secure the change necessary to it. But the point I want to make here is this, that the A-determinations which appear as predicates in such sentences as 'M is present/past/future' are for McTaggart fundamentally *relations* holding between events and a putative extra-temporal X. The predicative way of speaking is an equivalent shorthand in which these relations are represented indirectly as properties of events, on the grounds that "different relations [to the extra-temporal X) determine the other terms of these relations [i.e. events] as being past, present and future" (NE, 20). So McTaggart does not arbitrarily postulate timeless event-substances, he argues for the timelessness of the relatum to which events are bound by A-relations. Even if events should come out as timeless on this analysis (does an A-relation to something timeless require that the other relatum – the event – also be timeless?), the fact remains that the timelessness of events is not an initial input into it. Contrary to what is alleged by those who take the short way with his shorthand, McTaggart has produced a strong *argument* which in no way *relies* on detensing, though detensing might be justified by it.

I want to offer on a central matter an interpretation which in the end can be no more than an option of the soul, though justified perhaps by the fact that the rival interpretations are options of the soul no less. The central matter, which has been divisive without having been central to debate, is the status of 'now' in 'N is now future (present, past)'. By definition and uncontroversially, 'N is now future' means the same as 'N is future'. In other words, a given A-determination (e.g. future) of an event is trivially relativized to now, as well as non-trivially relativized to an A-determination of a different event. The relativization of A-determinations to now is trivial in the sense of non-truth-functional, in that if it is true that N is future, it is also true that N is now future. In this it differs from non-trivial relativization, in that if it is true that N is (now) future at a moment which is past, then it is false that N is (now) future in the non-trivial sense of 'now'. This trivial now ('clitic' now, perhaps, as in 'clitic pronoun') is wide open to interpretation the moment we move beyond the shared insights that (i) 'L is now future' means the same as 'L is future'; and (ii) that 'L is now future' does not mean the same as 'L is future *and* now'. Paradoxically or non-paradoxically, everything real is future and now and past, if the universe is tensed, but that is not a fact to which the now in question contributes.

If we go along with the idea that A-determinations are relativized to this trivial now, it follows that McTaggart's distinction, corresponding to the option between contradiction and regress, between absolute A-characterization ('M is present' *tout court*) and relativized A-characterization (M is present at a moment which is present . . . etc.) is too stark and should be replaced by the weaker distinction between trivial and non-trivial relativization. His distinction is "too sharply drawn", as one philosopher has noted, adding, I think wrongly:

> Where A-determinations are trivially relativized in this way they are just as tensed as where their relativization is non-trivial. Their distinc-

tive feature is not that they are detensed but that the tense is a present tense which is used contrastively in conjunction with other tenses.[8]

The revised (weaker) distinction is sound inasmuch as it draws attention to the fact that a now is intrinsic to, tacit within, 'M is present'. Whether this means that the now of 'M is now present' says when M is present is another matter. If it does, there is no need to go chasing down the regress opened up by the non-trivial relativization of 'present' to a meta-moment external to M. Saying that M is present at a moment which is present at a moment which . . . etc. would be saying no more and no less than that M is present: we would be inducing self-hypnosis with this tautological iteration of M's present moment, not vertigo from climbing up the strata of time.

But why should the fact that 'M is now present' means the same as 'M is present' *not* be taken as triggering the regress 'M is present at a meta-now which is present at . . . etc.'?[9] Talk of the "iterative implication"[10] of A-determinations cuts both ways, allows for iteration of A-determinations within the first-order series as well as throughout meta-series. The mere fact that 'M is present' means the same as 'M is now present' entails neither option. However, if the now is taken in the first way, as saying within the first-order series when M is present, then we must be dealing with a tensed now (as the quote claims) which reflexively designates *itself* as the unique tensed time when M has a given A-determination. For only then does the time when the A-determination holds not need to be externally designated by a meta-now. There would have to be a succession of these nows, the one designating itself as the unique present time when M is present and only present, the next as the unique present time when N is present and only present, and so on.

But is there really such a succession? Given 'M is now present' and 'N is now present', do the two 'nows' denote *different* present times, the one succeeding the other? I submit they do not, and that in general there exists no succession of trivial nows. I have been looking at a photograph from 1900 of a woman crossing a snow-covered square in a German city. I have the feeling, which old photographs are known to induce, of sheer incomprehension in the face of the fact that this woman is dead. Why make a mystery, as I mean to, out of her being dead, considering that there is nothing mysterious about the physical processes that lead to death? Because when I say to myself 'it is now when the photo is present beneath my gaze' and imagine the woman having said to herself 'it is now that snow is present beneath my gaze', I am convinced that the now in both cases is one and the same now. Not two successive now-tokens of the now-type that is essentially connected with the coming and going of events (in that for an event to be is for it to be-now). What is incomprehensible is that that woman should have exited that transtemporal (or purely temporal) now that remains the same through all empirically differentiated nows.

This now that is always the same is not the now we call 'current', not the space-bound now Heidegger calls "occurrent" (see pp. 192–193). Which is to say that it is not a time. And because it is not a time, it does not replicate the first-order series of occurrent nows, and so does not take us into the regress. But for the same reason, that it is not a time, it does not answer the question, 'When is M present and not also past and future?'

Anyone who still says that the trivial now is present-tensed will have to say how it relates to the non-trivial A-determinations in conjunction with which it is said to function; in particular, how it relates to non-trivial presentness, with which it seems to double up uncomfortably. It is no use referring to the facts of linguistic usage ("is used"), for these facts are conceptually opaque. What sort of contrast, for example, is the one alluded to in the quote, considering that the trivial now does not belong to the same system of contrasts that hold between the A-determinations? Each of these contrasts with the other two; whereas the trivial now would be a fourth A-determination rather than a dimension common to all three, if it contrasted with them in the same tensed way as they contrast with each other.

Analytical philosophers have systematically ignored the status of now as always the same and the problem of its relation to now as always different, presumably because 'now is always different and always the same' is deviant logic. I think the correct response here is 'so much the worse for logic'. For I maintain that the following is an irresistible intuition, as commonplace as can be: when I think today that it is *now* that rain (is) **now**, having thought yesterday that it is *now* that snow (is) **now**, the italicized 'nows' name the same now while the bold 'nows' name different nows. The first therefore differs from the second in a much more radical way (which we might call 'supra-dimensional') than a second-order time-series differs from the first-order time-series of the same (**now**) type.

The trivial now could not be a candidate for McTaggart's timeless entity X; for that entity is qualified by A-relations whereas the now in question is essentially unqualifiable. The latter is supra-dimensionally different from the empirical now in lacking differentiation through contrast, which is why it cannot say when an event possesses just one A-determination. But what if there existed a stratum of time that were both supra-dimensionally different from the series of empirical nows (tensed events) *and* were internally differentiated in some way deeper than tensed differentiation? A solution in this direction will be attempted in the last section of this chapter.

Against McTaggart's predicative model of tense

Substantial consensus prevails among defenders of tensed time that McTaggart's whole predicative model of tense is deeply suspect. Things (or whatever) that change are existing things, and it is *while they exist* that they change, whereas the picture of change set up by McTaggart's treatment of tense as a special case of predication ('M is present', etc.) suggests that things exist before they have begun to be ('M is future') and after they have ceased to be ('M is past'), so that their temporal existence seems to be a phase of their sempiternal existence. The unborn Napoleon waits in the wings to make his entry onto the stage of actuality, frets his hour thereupon, then moves into the wings on the other side, out of the spotlight of presentness but still supplying an existent subject for 'is dead'. This picture is to be avoided, the argument runs, by replacing the pseudo-predicates 'past', 'present' and 'future' by ordinary tensed verbs, which neutralize the suggestion of immortal subjecthood carried by the predicative model. This approach has been formalized by A.N. Prior, whose tense-logic is erected on the postulate, pictured by verbs and not by predicates, that there are no facts about entities not at present in existence. I turn now to the question of how, if at all, this affects McTaggart's argument.

Prior points out that all overt indications of the present tense are equivalent to the vacuous 'It is the case that...', read as standing in contrast with the non-vacuous phrases 'it has been the case that...', 'it will be the case that...' (plus the phrases for complex tenses) rather than in contrast to 'it is false that...'.[11] Thus, 'I am now eating my breakfast', says no more than 'It is the case that I am eating my breakfast', and even here the present tense is sufficiently conveyed by the unmodified root-form of the verb (by *eat*), so that the tense-indication 'am ...ing' is dispensable. The sheer verbality of finite verbs shouts out the present tense. Or to put the point in the material mode, an event is present in virtue of being an event, not in virtue of being an event *which is present*: "the presentness of an event is just its happening".[12]

Prior calls his theory "a no present theory",[13] which is somewhat misleading considering that he means just that *indications of* the present tense are superfluous. His main point about the present is that, far from there being none, it is "what we have with us all the time" and deserves to be called, more appropriately, "omnipresent".[14] An event's ceasing to be puts an end to the *event's* presentness, but not to presentness being involved in *our* relation to it. The present is what is always with us in two ways. First, pastness is past presentness, futurity future presentness. Second, presentness is, as we saw, *present* presentness, pastness is *present* pastness, futurity is *present* futurity.[15] Thus the past is *in toto* present past presentness, and so on. Presentness, in other words, flanks the tensing on both sides.

But all this gets us nowhere, in the absence of ontological elucidation from Prior. Is the present on the right-hand side of the tensing the same sort of beast as the presentness on the left-hand side? If it is (what I have denied), then it is a serial present and no sooner enters into a relation with past presentness than it is ousted from it by the next present. If it is different (as I have claimed), then there is more to the theory of time than the theory of tense. As to the presentness on the left-hand side, if the 'past' modifying it is construed truth-functionally as making 'M is present' false, then the chances of *our* relation to events continuing after their obliteration are blighted by the fact that there is nothing there to have a relation with.

'Past presentness', as understood by Prior, does in fact mean sheer privation of presentness, not a new sort of positive presentness (and similarly with future presentness). In order to ensure that past and future events enjoy no kind of positive being (that "there are no facts about them"), Prior proposes that non-present tense-modifications be construed as modifications governing the *whole* sentence, such that simple tense operators function in the manner of sentential adverbs. Thus 'P (I am eating my breakfast)' says that my eating my breakfast was a present truth but no longer is, while 'F (I am eating my breakfast)' says that my eating my breakfast will be a present truth, and 'FP (I am eating my breakfast)' that my eating my breakfast will have been a present truth. 'P (I am eating my breakfast) stands to the core 'I am eating my breakfast' as 'Allegedly I am eating my breakfast' does. All contingent truth is by this account present truth: to say that something is past (future) is to say that there were but no longer are (are not, but will be) *present* facts about it. And it is supposed to be an advantage of this verbal-cum-adverbial model of tense that it gets rid of the sempiternal subjects of mutually contradictory A-predicates, thus of McTaggart's contradiction.

The force of Prior's 'P' and 'F' is to consign presentness to one of "these two great species of unreality, the past and the future".[16] His tense-operators effect the *cancellation* of presentness without compensatory reinstatement of what is present in some positive ontological status derived from presentness. But this is to treat temporal negation as a special case of ordinary logical negation. The two kinds of negation are, however, crucially different, despite both having in common the properties of iterativeness and mediation of narrower by wider scopes. The cancellation of an affirmation is not equivalent to the denial (withholding) of an affirmation: the cancellation 'I have eaten my breakfast' is different from the negation 'I am not eating my breakfast'. And double negation gives us a positive, whereas two prefixed 'P's' do not give us the positive 'I am eating my breakfast', but 'It was the case that it was the case that I am eating my breakfast'. For all that they are named 'P' and 'F' (etc.), Prior's tense operators void the universe of the bracketed propositional contents just as much as 'false' does. One result of this voiding is that on Prior's account we cannot significantly say that people are dead. Where 'P' is construed as a cancellation without ontological reinstatement, 'P (Napoleon lives)' means that upon quitting the scene Napoleon ceases to have entered it – he joins the ranks, not of the dead, but of the unborn.[17]

I shall not go into Prior's analysis of the future, except to remark that his statement, depressing for weather forecasters, that "although I can say *now* that snow was future yesterday, I could not have said *yesterday* that it was future" requires us to believe that the future is *originally modified*, a past future – too high a price to pay for avoiding future facts. The problem is as follows: on the one hand, it seems inconceivable that present facts should introduce us to the concept of futurity, via past futurity. On the other, the alternative seems to involve treating the future as a set of real events waiting in the wings. But that is not the alternative. It is true that the snow falling today was not future yesterday; but that is not because yesterday, when snow was not actual, I could not have legitimately said that snow was future. It is because the snow I could have legitimately said was future was *some* instance of a general snowy event-type, whereas the snow actually falling today is a particular, unpredictable, token of that event-type. *This* snow, the snow that is falling now, was not future yesterday, just because this snow is essentially snow in the grip of happening and no other. But there is no chance of accommodating the generality of predictions, or making sense out of talk of the simple future (or past) within an ontology which reduces what-is to what is happening. This ontological economy is not slimming – it starves us of our concepts.

Prior's difficulties over pastness and futurity are predictable from his atomistic postulate that there are no facts about individuals not at present in existence. Behind his verbal-cum-adverbial recasting of the predicative (adjectival) model of tense lies the Augustinian doctrine that the whole of reality is exhausted by what is going on now, that time is a succession of windowless presents – a doctrine which rules out change and replaces it by pseudo succession (succession not presupposing change, 'false' masquerading as 'past' and 'future').

Strictly speaking, McTaggart's paradox cannot even be formulated on this atomistic postulate (which is why our atomistic common sense is not puzzled). There is no *problem* about an event being present, past and future once the future and past have been dismissed from existence by tense-operators. That is the unavowed, more or less visible, rockbottom of all the objections levelled against McTaggart from the standpoint of tensed theory. However, even if we make this atomistic doctrine a gift of tenable concepts of pastness and futurity, the familiar difficulties still arise. Any tensed sentence or proposition must be true at some times and false at others, otherwise it would not be tensed. To say in tensed terms when a tensed sentence is true we should need a tensed meta-sentence, and so on endlessly. Sentences are no better at not being both true and false than events are at not being present past and future.[18] We might try pinning down a time when the tensed sentence 'The World Trade Center was attacked yesterday' is true by translating that sentence into the token-reflexive sentence, "The attack on the World Trade Center is tenselessly earlier by one day than the utterance, on 12.9.01, 'The World Trade Center was attacked yesterday'". Suppose (but just for a moment) that this tenseless token-reflexive sentence does translate the meaning of the tensed sentence: then since the former is true at all times if ever true, it does say when 'The World Trade Center was attacked yesterday' is true. But now we would have solved the problem of saying when tensed sentences are true by tenseless means which do nothing to alleviate the contradictoriness of tense.

The formal differences between models of tense are philosophically irrelevant whenever they are interpreted, as they have been, in the light of the same metaphysical decision in favour of temporal atomism. The models then differ only in form and in how overtly they treat 'past' and 'future' as special cases of categorical negation. Thus it has been proposed that instead of switching from predicates to tensed verbs, we might remain with predicates, but interpret them as a set of correlative operators logically akin to the "alethic operators" 'true' and 'false'. From this perspective it has been claimed that when McTaggart says that an event which is past is also future and present he fails to see that

> the futurity and presentness of an event that is past . . . is mediated by the pastness of that futurity or presentness; and by this mediation by a pastness of a wider scope it is thereby *cancelled*, because the wider pastness is itself unmediated.[19]

But surely the fact of the matter is rather that *no* futurity and presentness can *have* pertained to an event that is past once the mediation of futurity and presentness by unmediated pastness has, as here again, been construed as the *cancellation* of the event's futurity and presentness. If the temporal predicates work like alethic operators, then 'past' converts 'present' to 'non-existent', not to 'was present': it simply cancels 'present', as God might erase a mistake from his creation.

The task of conceiving 'tensed' change non-atomistically is the task of conceiving it as somehow at one with the stasis in time. For abstracted from the stasis, the dynamism is atomized, while abstracted from the dynamism the stasis is spatialized – as I shall now try to show.

The objection from tenseless theory

This line of objection agrees with McTaggart that tensed change is self-contradictory, but denies that this entails that there are no temporal relations (that time is unreal). It is argued against McTaggart that the relations of the B-series entail change, thus qualify as temporal relations, independently of the impossible A-series. Time is saved by the B-series, though there are in reality no tensed facts.[20]

The issue here is not the hotly debated one whether A-statements are reducible to B-statements, by translation of the former into the latter or, should such translations prove impossible (as the 'new detensers' have argued), by showing that A-statements nevertheless have tenseless token-reflexive truth conditions. (In the latter case the meaning of 'It snowed last week' would be *given* by the tenseless token-reflexive sentence 'Snow is earlier by a week than the token 'It snowed last week' uttered at (e.g.) 13.7.04', despite the tensed-token not *having* the meaning of that B-token.[21]) The reduction would be a lost labour if it is true, as McTaggart claims, that B-relations by themselves, ungrounded in A-relations, fail to qualify as temporal relations in not involving change. Whether they do or not is the issue that concerns us here. Does the fact of a poker being hot on

Monday and cold on Tuesday involve a change once it has been denied that the poker changes in the sense of *becoming* cold? It does according to the definition of change on which all B-theory relies, namely Russell's:

> change is the difference in respect of truth and falsehood between a proposition concerning an entity at time *T*, and a proposition concerning the same entity and the time T_i, providing that these propositions differ only by the fact that *T* occurs in the one where T_i occurs in the other.[22]

Notice that no A-terms occur in this defintion. We are not to think of the events of the B-series as being ordered by the A-order of their becoming, such that an apple would be green earlier than it is red only because it *turned* red after becoming green. Change is reduced to the fact of a thing having different properties at different tenseless times. To Russell's definition, McTaggart replied:

> I am unable to agree with Mr Russell. I should indeed admit that, when two such propositions were respectively true and false, there would be a change. But I maintain that there can be no change without an A-series. If, with Mr Russell, we reject the A-series, it seems to me that change goes with it, and that therefore time, for which change is essential, goes too.
>
> (NE, 14)

This confrontation between Russell and McTaggart is the heart of the matter. Before coming to it, however, it is worth noting that McTaggart is quite happy to go along with Russell making things, rather than events, the subjects of change: "Russell looks for change, not in the events of the time-series, but in the entity to which these events happen, or of which they are states" (NE, 14).[23] That it be a *poker* (a thing) that changes from hot to cold (rather than the event of it being hot from present to past)

> makes no difference to the qualities of the poker. It is always a quality of the poker that it is one which is hot on that particular Monday. And it is always a quality of that poker that it is one which is not hot at any other time.
>
> (NE, 14–15)

McTaggart is saying here that the existence of changeless *facts* about the poker ("it is always a quality of the poker that . . .") means that the *poker* (a thing) cannot change, whereas Russell and his fellow tenseless theorists claim that, though the facts about the poker never change, this is no impediment to the poker changing: it changes by dint of being hot at one tenseless time (a Monday) and cold at another (a Tuesday, say). So the issue whether a tenseless B-series can be a time-series comes down to whether it can accommodate change, and this in turn comes down to whether things can change despite the facts about them never changing. Tenseless theorists assert that they can, McTaggart merely counter-asserts (as we have just seen). But I want to return an answer that favours McTaggart.

I take the double aspect of change – a thing changing consistently with the facts about it never changing – to be so rudimentary to our understanding of time-order as to merit the status of a condition of adequacy on any theory of serial (ordered) time. In the case of tensed time the distinction is clear: having *become* cold, the fact of the poker being hot before it is cold *remains* changelessly in force. The tenseless thesis, however, must be able to offer its own version of the distinction, if it is to make good its claim that changeless facts are no impediment to changing things. But it cannot. For what is it for the poker to change from hot to cold, on tenseless theory? It is for it to be at all tenseless times true that the poker is hot on a Monday and cold on a Tuesday. And what is it for there to be changeless facts about the poker? It is for the very same truth to hold. But the same truth cannot distinguish between the two aspects of change. So either a thing changes without there being changeless facts about that change, in which case we have no fixed time-*order*; or else there are changeless facts involving no changing thing, in which case we have no *time*-order, as McTaggart says.

Russell's definition devotes the mentioned tenseless truth to proposing an order that is not an order of change (not a *time*-order) in that it meticulously, almost comically, misses out the metaphysical event of change, which is the *making* of the difference pictured (or whatever) as the difference between the propositions he mentions. Change happens 'in between' the world being one way and then being another way, it cannot be *defined* as the world being one way and then another, though that difference is surely a *criterion* of change. It is true that my poker changing from hot to cold *entails* that the proposition 'My poker is hot at t' is true and that the proposition 'My poker is hot at t_1' is false. It is true, in other words, that a change generates a succession, which is criterial for it. The philosophy of time, however, is a museum of pseudo successions – of orders whose claim to be successive is unsupported by a sustainable concept of change. The atomistic conception of tense was one example. I want to suggest that Russell's definition yields another.

A poker that is hot at one time and cold at another is a poker that is hot and cold. But this involves no contradiction when the times are understood as B-times (e.g. calendar dates), since anything that has to anything else the relation 'earlier/later than' always has it and it alone (whereas anything that is present is also past and future): any particular situation S that is ever earlier than another particular situation SS is not also a situation that is later than SS. But we might ask where these B-times come from. Are they really times i.e. successively ordered? On A-theory the answer to the question where invariant time-relations come from is clear enough: they are generated by becoming, by the change of events from future to present to past; if event

B is ever just-past relative to present A, then B remains forever thereafter (at all future nows) just-past relative to A. But where do *Russell's* tenseless B-times comes from? They cannot be generated by changes, given that he defines changes in terms of times and not times in terms of change. I submit that the times to which Russell helps himself are the series of natural numbers in drag: the series of numerical subscripts is contradicted by '*t*', inasmuch as '*t*' implies succession (a time is essentially earlier and later than other times) while numbers are not objectively successive – they 'come after each other' only in the subjective process of counting. I conclude, then, that B-theory (the theory that denies that temporal relations become) fails to secure a coherent concept of invariant *time*-relations, distinct from spatio-cum-mathematical relations exclusive of change. B-theory's invariant relations are pseudo successions, numerical relations into which succession has been smuggled back *in memoriam* of their origin in the A-series.

The prospect of a solution

Having concluded that no genuine dynamic succession is possible where tensed change is construed in abstraction from the stasis in time (i.e. as cancellation), we now find that no genuine invariant succession (temporal stasis) is possible where this is construed in abstraction from the dynamism of time: construed, that is, in terms of tenseless theory. Conversely, the problem with the interpretation of temporal dynamism as analytical philosophy's truth-functionally-conceived 'tense' is that it turns out, in whatever logico-linguistic dressing, to exclude the stasis or 'glue' in time to the point of atomization. In short, *tenseless theory spatializes temporal stasis, while tensed theory atomizes temporal dynamism.* These symmetrical errors of spatialization and atomization (varieties of pseudo successions) suggest that McTaggart's as yet unresolved paradox thrives on the dichotomization of two aspects of time which really belong together: a dynamism harbouring a stasis, as distinct from an A-time conceptually distinct from a B-time. For it is a sound Hegelian insight that abstractions do, sooner or later, lead to contradictions. I now want to pursue this thought in the context of a Husserlian theoretical framework deconstructive of the dynamic vs static dichotomy.

Towards a Husserlian solution

We are familiar with two species of serial ordering 'next to', which generates the spatial series, and 'becoming before', which generates the A-series. The spatial series is homogenetic, by which I mean that its terms do not have 'birthdays', whereas any dynamic conception of serial time acknowledges that it is heterogenetic, that is, comprised of events having different birthdays (coming to pass at different presents). The concept of heterogenetically serial time, as I propose it here, says no more than that. It does not say, additionally, that every birthday *inflicts a death* (or every set of birthdays a set of deaths). That doctrine, that every nascent present event kills off its antecedent, is the doctrine of tensed time, which is a particular interpretation of the conception of heterogenetic time-series. What makes Husserl and McTaggart significantly comparable is that their conceptions of dynamic time fall as variants under the general concept of heterogenetic time-series. Whereas the times proposed by Bergson, Heidegger and Levinas are not even serial (ordered), let alone heterogenetically serial.

Let it be borne in mind that the endorsement of Husserl that will emerge in the confrontation with McTaggart will not imply that I agree with his pervasive tacit assumption that time is always, or even usually, experienced as a heterogenetic series. I proceed on the basis that Husserl's time-theory retains its value as a phenomenology of the experience of *serial* time, with no suggestion that this exhausts the phenomenology of time, but with the strong suggestion that much of its value lies in its capacity to intervene therapeutically into the familiar paradoxes associated the analytical conceptions of serial time.

Summarizing how invariant time-object-points emerge out of the flow of retentions, Husserl affirms that "[t]ime is fixed and yet time flows" (ZB, 64). This suggests that he acknowledges a distinction *apparently* corresponding to the tenseless and tensed times of analytical philosophy of time, although no equivalents of these adjectives are to be found in his texts. My business with McTaggart requires that we find out just how Husserl's interpretation of the relation of temporal dynamism to temporal stasis compares with the analytical theorization of that relation in terms of tensed A-time and tenseless B-time.

It will help bring out the specificity and originality of Husserl's position if we offset it against C.D. Broad's 'analytical' account of the order of time (thus of time's direction). Having asked how the static time-order differs from other static orders, such as the spatial and the numerical, Broad points out that the difference cannot be grounded in different logical properties of the generating relation in either case; for it is a fact that a non-temporal relation such as *larger than* and the temporal *earlier than* are both transitive, asymmetric and irreflexive. Yet the time-series possesses an intrinsic direction lacking from the number-series and a one-dimensional spatial order: direction can only be introduced into these orders extrinsically (by e.g. counting from left to right). On the basis of these considerations Broad concludes that "the intrinsic sense of a series of events in Time is essentially bound up with the distinction between past, present and future. A precedes B because A is past when B is present."[24] Invariant relations of precession therefore presuppose the facts of tensed becoming: there can be no *independent* B-series (or plurality of B-series, if we want to bring different frames of reference into play), contrary to what is implied by those philosophers who uphold the autonomous reality of the B-series while assigning tensed time to unreality or merely subjective reality. Husserl and Broad agree that invariant time-relations exist only as a property of a series whose terms become in the now. So in calling these relations 'invariant' we cannot mean that they are tenseless, if 'tenseless' means 'pertaining to a series not generated by now'. For there exists no such series, according to Husserl and Broad. The stasis of time is for both these philosophers a function of its dynamism. What is excluded by time's fixity is not time's flow, but rather fluctuation in the order *of flow*. But what we want to pin down here is where Husserl differs from Broad.

I think most people would agree that Broad's position is *commonsensical*: 'Of course event A can be invariantly earlier than event B only if A has happened when B is happening (only if A is past when B is present).' But I do not think this commonplace is *intuitively self-evident*, whereas it is a merit of Husserl's account of the priority of temporal dynamism over temporal stasis that it *shows*, in terms of connections of eidetic meaning, how the flow of time *yields* the stasis of time as a property of the flow itself, in the way we have seen – intuitively *seen* (see above, p. 57). If I may risk exploiting an etymological relationship with no insinuation that it houses Being: the 'still' of 'still the same time-relation throughout the recession from the actual present' is a stilling of dynamism, and this wondrous stilling of dynamism *is* temporal stasis. The stasis of time can no more be separated from the dynamism of time than the immobility of the fountain from its mobility. By contrast with this, the notion of presupposition on which Broad rests his case for the priority of A-time is intuitively opaque, for all that it is common-sensical. And that opacity is precisely Husserl's focus, inasmuch as he makes "the main theme of the theory of time-consciousness [the elucidation of] how the objectivity of the time-point gets constituted within the constant flow of time-consciousness" (ZB, 428). The procedure of constitution here may be seen as pre-empting problems which arise from the language of presupposition. If one series presupposes the other, does that not mean that there are two series? And if there are two, is it not a strange coincidence that the order in which they arrange events and the measure of temporal distance between them are the same in each? And why do we all believe that there is just one series?[25]

Husserl's position is that there is just one time with a dynamic and static aspect and that (though here I speak for him) the tensed vs tenseless dichotomy is obtained by abstraction of these aspects. At any rate, Husserl's 'flow' and 'rigidity', being aspects of just one time-series, do not respectively match the tensed and tenseless times of analytical philosophy. The unity of dynamic and static time is thinkable only from the standpoint of consciousness, thus not thinkable from Broad's standpoint of the time of events considered without reference to consciousness. As I have said, the 'still' that denotes stasis is a stilling of dynamism: a past phase of consciousness is without contradiction in play together with (extensionally all-at-once-with) the present and future phases. What are in play together, what constitutes the original stasis in time, are invariantly ordered variations of modes-of-being of phases. A past phase of consciousness is a modal transformation, designatable by the relatively positive word *was*, of a present phase of consciousness; whereas an event (-phase) that is past has undergone the non-transformative change we refer to by the relatively negative expression *no longer*. The former does, while the latter does not, survive the present in a new mode-of-being. Putting this asymmetry differently: a past event is not what a present event was, whereas a past phase of consciousness is forever what a present phase of consciousness was.

It is only if *no longer* is somehow different from *was* and grounded in it (i.e. only if the time of events is grounded in the time of consciousness) that we can hope to escape the atomism of tense-theory – its disregard for the difference between its starting to *have been* a present fact that E happens and E never having exited the domain of the Great Unhappened. It cannot be answered that there are events (imaginary ones like the sun exploding) of which 'is a present fact' has never been true, and that the function of the expressions *was* and *no longer* (assimilated by tense theory) is precisely to distinguish from such events the subset of events of which 'is a present fact' was, but no longer is, true. This rejoinder does no more than plead that our use of words shows that we *mean* to distinguish between categorical and temporal negation. It does not say what we mean by the distinction we mean to make: in effect, it leaves all negation categorical.

The asymmetry between the time of consciousness and the time of events seems to disable what might have looked like an answer to McTaggart. We cannot say 'Look, the phases of absolute consciousness are in fact together without contradiction, so you have no paradox.' For this does not do away the paradox at the level at which McTaggart formulates it, which is the level of events and their presentness pastness and futurity. Consciousness escapes paradox precisely because it is not, as we have seen, an event with present past and future properties. The existence of the asymmetry does not, however, rule out appealing to consciousness in attempting a solution at the appropriate level of events. Indeed, any Husserlian solution would have to appeal to consciousness, since the time of events is constituted by it; which is why the solution I shall broach in a moment will be set within an idealist framework. This might look like a shortcoming, inasmuch as a solution essentially involving consciousness will not be able to take account of events (such as McTaggart's death of Queen Anne) that nobody is in a position to remember.[26] Against this should be set, however, the artificiality of McTaggart's treatment of events as occurring for nobody: an event that goes through tensed changes for nobody simply puts the limp on the other, realist, foot.

Recall, now, that the problem was how an event *can* be past, given that it is also present and future and that these temporal determinations exclude each other. What we are looking for is a non-regressive factor of separation in virtue of which past, present and future do not qualify an event conjointly.

In what is perhaps the only statement in the whole of ZB evocative of

McTaggart, Husserl offers this enigmatic answer: "Identically the same thing can indeed be both now and past, but only by enduring between the past and now", having conceded the truism that "now and past exclude one another" (ZB, 318). Husserl evidently means that the fact of a thing enduring puts temporal distance between its being present and its being past. This of course invites the rejoinder that the requisite temporal distance could not consist in anything other than the thing's presentness and pastness being the case at different meta-moments whose temporal difference cannot be consistently stated. It is obvious, then, that if Husserl's statement is to escape McTaggart's regress, the mentioned "endured" must refer to something quite different from tensed change from present to past.

To make headway from here we must first get clear about the Husserlian-phenomenological situation(s) in which the attribution to an event of pasteness and presentness applies. Now, a contradiction, such as the one alleged between 'M is present' and 'M is past' is a potential property only of *judgments*, not also of sensory experience. The present phase of *consciousness* does not contradict the past phases with which it is 'together' (extensionally all-at-once) because these phases, though mutually exclusive in temporal meaning ("time-value", ZB, 285), are not properties predicated in judgment of a thing or event. For as we have seen, the absolute flow is not a 'thing which flows', not something assimilable to the subject-predicate structure of *judgment*: its 'enduring' involves no changing of a something from present to past in the ordinary predicative sense of these temporal adjectives. It is a sensing, having in common with 'ordinary' sensations (of e.g. coloured surfaces) that its constituents cannot contradict each other. I cannot without contradiction say (judge) in the same breath of the same thing 'it is white all over and black all over'. But if the thing in question is a block of white marble over which a dark shadow has fallen, I do not see 'white conflicting with black', but rather the composite colour grey.[27] In short, contradictions require judgments; judgments require subjects of prediction; the absolute flow involves no subject and no temporal predicates; whence it follows that it cannot host McTaggart's contradiction.

The question is now whether the appearance of contradiction can arise at the level of the *time-objects* constituted by the absolute flow. Now, 'past', 'present' and 'future' are words which Husserl is prepared to apply to object-points and objects (to the shrill phase of a whistle blast and to the whistle blast as a whole). They, and the co-ordinate acts, possess "a certain determinability with respect to before and after" (ZB, 236). I take Husserl to mean that at this level 'past' and 'future' refer to *horizons*, which are not yet fully-fledged relations of predication. The horizonal 'past' and 'future' are together as *apperceived*, whereas the phases of the original absolute consciousness comprise a *felt* togetherness. Apperception of the horizonal past and future of objects is not yet a predicative or judgmental apprehension, though it is nearer to this than original time-consciousness. This would seem to be confirmed by Husserl's earlier-considered contention that to be fixed as "properly objective" the time-order constituted in the original field must be reinforced by the reiterations of memory: the 'topping-up' by memory would effect the passage from a not-yet predicative ("a certain") determinability by pastness and futurity to a fully-fledged one. In short, this intermediary level (intending time-objects as set within their time-horizons) is still inhospitable to McTaggart's contradiction because we do not yet have subjects of predication or (equivalently) propositional contents.

That in our immediate perceptual relation to time-objects we do not *judge* them to be past, present and future would seeem to be confirmed by a passage of *Ideen II* where Husserl speaks of the products of original syntheses as purely sensory objects or "ur-objects to which all possible objects refer back in virtue of their phenomenological constitution".[28] Such ur-objects are neither objectivized in space nor posited as subjects of judgments. We can hear a sound without judging '*That thing* (located in space) is making a noise': "[i]t is evident that such a sound-datum could be constituted without a spatial apprehension even being performed ... A sound is conceivable which lacks all spatial apprehension".[29] There is no "subject-positing" (*Subjektsetzung*) here, no performance of a "categorical synthesis" by which a thing would be linked in judgment to its changing or persisting states: "with the pure [pre-objectivated] sense-datum we come up against a pre-givenness that lies even before the constitution of an object as an object". As apprehended in this way, time-things are purely sensory products of "the ultimate syntheses [of time-consciousness] which comes before any thesis".[30] By this account, the '*M*' of the McTaggartian statement "*M* is past" would be the 'thesis' of which 'past' is predicated in a categorical synthesis, while the ur-object would be the temporal unity pre-given (pre-constituted) as "substratum" to the categorical constitution of the thesis. Because it is a product of pre-categorical syntheses, the ur-object is already a tissue of time before the new lease of temporal life it acquires as a thesis or subject of temporal predicates.

But the objective time fully constituted by memory does have predicative structure: "[t]emporal objectivity is produced in the subjective temporal flow, and it is essential to temporal objectivity to be identifiable in recollections and as such be *the subject of identical predicates.* "(ZB, 108; emphasis added). The position now is as follows: Husserl would accept, as judgments of memory: 'M is present', 'M is past', 'M is future'. But he would deny that we have here a contradiction. How can he? The solution to be proposed will rely on the notion of the containment of the pre-predicative time-track within the predicative time-track (otherwise roughly expressible as the groundedness of *no longer* in *was).*

If it is true that the predicative understanding of time arises only through memory, it follows that the temporal predicates apply within a dimension of pastness: the pastness of M is its having been present and future in the eye of a new present which is the remembering act N. ('M *is* future', in the sense of existent and qualified by futurity, is clearly illegitimate.) Not that this removes the contradiction, if there is one, since M's having been present and future when it is past is still the problem how M can *be* past and (have) *be*(en) present and (have) *be*(en) future. But it does place the problem within the perspective of memory.

What appears to a present remembering act N is event M as presumptively having been present and future autonomously, that is, prior to being represented as past-present and past-future. M is past in its 'rememberedness', present and future as having autonomously preceded its rememberedness. These predicative time-determinations, constituted by memory, are mutually exclusive. Underlying them, however, is a continuity, the continuity referred to in Husserl's earlier-quoted (p. 132) remark that, if what is remembered existed prior to the remembering act, then necessarily it belongs to the same stream of consciousness as that to which the remembering act itself belongs. Or again: "It is . . . evident that two such streams, which have an experience in common [e.g. perceiving X and remembering perceiving X] enter as parts into the unity of one encompassing stream".[31] This is the continuity evidenced in the 'presence' of what is remembered, the presence contested by the absence of presence, or discontinuity with the present, which pertains essentially to the thetic character of remembering acts. This tension is not thought of as 'a commingling of past and present', but rather as an undergirding of a relatively discontinuous time-track by an absolutely continuous one.

Which is to say that: as belonging to one and the same stream, the predicative A-determinations (present, past and future) are in their pre-predicative micro-structure extensionally-all-at-once, they are *together as mutually differentiated* phases of the stream, which phases provide a non-regressive principle of pre-predicative separation (and unification) in virtue of which the same event can be present, past and future disjointly but without the disjunctions depending on past, present and future. There is no vicious regress here because the phases of the time-stream, though differentiated, are not differentiated as present, past and future (as we have seen). In short, we have no contradiction because the attributions of the predicative A-determinations correspond to, and are interpreted back into, differentiated sections of the pre-predicative continuous time-stream.

In explanation of 'interpreted back into' let me risk this image: think of Husserl's time-stream as a length of string and of predicative A-determinations as knots tied in it by memory. Though the mutually exclusive time-knots stand out from the string, they are nothing but resolvable complexifications of it. An equivalent image would be river-borne lumps of ice poised between congealment (predicative status) and dissolution back into the (pre-predicative) flow.

It seems to me that this explanation fits with what goes on in the mind of any philosophically unprimed person encountering the statement that an event is present, past and future. We do not naturally jib at it: not because we fail to think the matter through, but because we have got the better of paradox from the start. We naturally maintain our poise between ('flicker' between) grasping events as belonging to either track, never running them on the one to the exclusion of the other. McTaggart's strategy consists essentially in upsetting that poised attitude, in portraying predicative time in abstraction from the pre-predicative time which saves it from contradiction. His tenseless and tensed times (the B-series and A-series) are obtained by abstracting from each other the complementary tracks of the memorial time-series: *no longer* from *was*. Concretely considered, the 'simultaneity' of extensional all-at-onceness is the simultaneity requisite for succession, the simultaneity *of* succession, the simultaneity that makes succession something different from repeated cancellation and replacement. The mentioned abstraction perverts this to spatial simultaneity; the cohesion of extensional continuity is tightened to the limit of timelessness, turned into the becomingless stasis misnamed tenseless *time*; while on the other side, the successional separations between events that are due to their differential A-determinations are pushed to the opposite limit of absolute discontinuity, fragmenting predicative time into an atomistic series of cancellations and replacements (tensed 'time'). With these abstracta in his hands McTaggart erects his paradox, irrefutable, granted its basis in abstraction, proving only that abstraction leads to contradiction and thence to paradox.

6 Is phenomenological time real time?

An outline of the problem

It is a characteristic doctrine of phenomenology that reflective analysis of pre-reflective experience elucidates the founding and unrevisable meanings of concepts which also have a life at founded levels, in philosophical and scientific languages. The following remarks by Merleau-Ponty are emblematic of mainstream phenomenology since the later writings of Husserl:

> Everything I know about the world, even through science, is known to me on the basis of a view that is mine or of an experience of the world without which the symbols of science would mean nothing. The whole universe of science is constructed on the lived world.[1]

It follows that questions as to what is meant by 'real time' (or 'real space' or 'real world') should not be referred to some post-experiential court of arbitration, such as science or logico-linguistic analysis. Phenomenological writings on time would, indeed, be only of psychological interest if we accepted the claims of those scientifically-minded philosophers who inform us that only fundamental physics can 'tell us what real time is' – not just what time is in physics but *per se* – and that the physical concept of time must supersede the testimony of time-experience. My aim in this chapter is to see what resources can be mustered within phenomenology to make the best case for its claim to be dealing with real time.

The immediately relevant background consideration here is that, with the exception of Bergson, all the phenomenologists take for granted Kant's doctrine that time does not inhere in nature, however far their views on time otherwise diverge from Kant's. Thus, Heidegger:

> The movements of nature which we determine in spatio-temporal terms do not elapse 'in time' . . . they are in themselves completely time-free; they are only encountered 'in time' inasmuch as their being as pure nature is disclosed. They are encountered in the time that we ourselves are.[2]

Heidegger is saying that natural things are temporal, not in themselves, but only in so far as they are disclosed through human temporality as belonging to our (Dasein's) world. Likewise Sartre: "The in-itself does not enjoy temporality precisely because it is in-itself and because temporality is the unitary mode of being of a being which is perpetually for-itself at a distance from itself."[3]

Opposed to this Kantian doctrine is realism about time, scientific and metaphysical. That temporal realism can be metaphysical as well as scientific – that it is not wedded to a quantitative conception of time and a physical conception of reality – is shown by the case of Bergson, who conceives time as radically opposed to quantity and as the basic stuff of a universe that is ultimately spiritual. Whatever Bergson says (and he says mutually inconsistent things about the cognitive value of science), his protest against quantitative time implies an instrumentalist view of scientific theories, such that they are to be interpreted as devices for facilitating predictions or as supplying a convenient framework for mathematical reasoning. Not as making claims that could be true in some correspondence sense. Time really inheres in nature, though not as a quantity. It was Bergson's stance against quantitative time, rather than his unKantian defence of time-in-nature, that influenced the French phenomenologists.[4]

Bergson, however, works with a unitary concept of time, whereas we have seen that the phenomenologists 'proper' distinguish between different temporalities, one of which they identify as fundamental and the other(s) as derived. This enables Heidegger to reject Bergson's unqualified proposition that quantitative time is a fiction and argue that, although fundamental time is not quantitative, the Aristotelian time derived from it is both quantitative and real. By 'real', however, Heidegger means 'genuine', non-spatialized: he does not mean 'inhering in a mind-independent nature'. Quantitative time would be true of the natural world *as a phenomenon disclosed* (Husserl would say "constituted") *within the human world*; it would be a founded mode of Dasein's understanding of *its* world as natural, which means that any reconciliation of pre-scientific and scientific time would have to happen on phenomenology's terms – as a tracing of the shifting meanings of concepts 'from the bottom upwards'. Phenomenological writings post-Bergson are in fact strewn with cursory programmatic pronouncements, made under the aegis of Husserl's project of "genesis of meaning", to the effect that the basic theoretical concepts of science could in principle be shown to be intentionally or hermeneutically dependent on lower-level experienced meanings fixed unrevisably in immediate experience: the meanings of the scientific concepts "send us back" to grounding baptismal meanings which are given in experience and of which the theoretical meanings would be preserving elaborations. Thus the baptismal sense of 'time' would be preserved throughout its scientific reconceptions, so that we are never dealing with two incommensurable senses which might be *found out* to have the same reference, contrary to the case with contingent identity-statements like 'the Evening Star is the Morning Star'(it is hard to imagine what this 'finding out' could mean in the case of time). John Findlay, for one, assumes the feasibility of this sort of project when he says that "the lines of the map of relativistic time are still his [Husserl's] lines of flux and becoming which have, in some way, been frozen into timelessness".[5] If only we knew in what way. And Merleau-Ponty has the same project in mind when he laments that the logical positivism which took philosophical charge of the new physics felt no need for a non-deceiving God, or a new philosopher who might ground Einstein's physics in experience, as Kant had grounded Newton's:

> classical science ... expected through its constructions to get back to the world. For this reason classical science felt obliged to seek a transcendent [Cartesian] or transcendental [Kantian] foundation for its operations. Today we find ... in a widely prevalent philosophy of the sciences an entirely new approach. Constructive scientific activities see and represent themselves as autonomous.[6]

Let us be clear that the sort of rapprochement envisaged by Merleau-Ponty could not be in the style of the integrated image recently mooted by Thomas Nagel (following Wilfred Sellars). The ultimate aim of this latter approach is to explain (the time of) subjectivity *non-reductively* as part of the real world, which means opening the prospect of the ontological honours turning out to be fairly evenly distributed as between scientific and pre-scientific understanding. The world of which (the time of) subjectivity would be a part would be neither wholly physical nor wholly mental. The task, as Nagel sees it, is to "go beyond the distinction between appearance and reality by including the existence of appearances in an elaborated reality. Nothing will then be left outside":[7] subjectivity will cease to appear as an "nomological dangler" (J.J.C. Smart) ripe for physicalist reduction in the name of the imperialist positivist principle of the Unity of Science.[8] Nagel's expanded reality (expanded beyond the physical) is what we glimpse when we reflect that 'after all, the first-person view of subjectivity is part of reality, for we cannot say that it is non-existent'. Now the reason why this particular conception of rapprochement would have been uncongenial to the phenomenologists is that it conflicts with their principle of the semantic dependency of the operation of high-level concepts on the baptismal meanings acquired in the human world – the principle of rapprochement from the bottom upwards.[9] In Nagel's picture, the statements of physics would be true of nature *autonomously* of truths about the human world: though the former would not replace the latter, they would not depend semantically on it. Nagel's style of anti-idealist rapprochement has no more in common with the phenomenological than the desire to be kind to pre-scientific experience.

The position represented by Nagel accepts scientific realism (the view that the theories of science can be true) while rejecting the 'replacement thesis' – the claim that first-person experiential statements can be true only once translated into the vocabulary of physics, thus voided of their first-person meaning. Though all replacement theorists are necessarily scientific realists, not all scientific realists are replacement theorists. For some realists deny the principle (not just the feasibility) of physicalist translation: some truths are irreducibly experiential. The replacement thesis is defined not by the realism it implies (as against instrumentalism), but by its metaphysical assumption that the real is exhausted by the physically real; whence it follows that the *only* empirical statements capable of being true are those of science, and that these ought to replace pre-scientific (phenomenological or other) statements in any description of 'what really is'.[10] The metaphysical assumption involved here begs a justification that would also have to be metaphysical, to the extent that it would have to come from outside science. But no such justification is offered in the scientifically-informed writings on time penned from the standpoint of relativity physics by some well-known American naturalist 'replacers'.

During most of the twentieth century the replacement thesis operated in a positive way, replacing 'subjective' tensed judgments with tenseless ones deemed to be consistent with the physics of Einstein. More recently, however, the prospect of a negative application of the thesis has been envisaged, following avowedly tentative scientific arguments for the unqualified unreality of time. The suggestion is that because time is about to "cease to have a role in the foundations of physics",[11] we had better start replacing our temporal ways of thinking about the world by timeless ways, and that "timeless principles will explain why we *do* feel that time flows"[12] (just as causal stories used to be offered to explain why we experience really tenseless time as tensed). The underlying metaphysical premiss here is the same as in the positive application, namely, that the real is exhausted by the physical, or equivalently, that either time inheres in physical reality or time is unreal. The difference between the old and the new replacers is just that to the question, 'Does time inhere in physical reality?', the first answer with a qualified 'yes' and the second with an unqualified 'no'. Neither explain why what 'we *do* [in reality?] feel' does not count as part of reality. (In this chapter I shall proceed on the premise that notions of 'real time' still regulate debates in the philosophy of time, much as a prize regulates a competition.)

There are, then, two ways of construing the problem of reconciling the two ways of talking about time: *à la* Husserl/Merleau-Ponty, and *à la* Nagel. And there have been two ways of getting rid of it: either by opting for instrumentalism and thus abstaining from verbal interpretations of the mathematical formalism, or by voiding first-person experience of cognitive significance. On the latter option, taken by the replacement thesis, no problem of rapprochement remains because the competing experiential truth has been translated out. The sole truth about time is enshrined in the verbal interpretations of the formalism, whereby ordinary words are re-baptised with extraordinary meanings.

But a problem does remain. For unless the validity of our pre-scientific understanding of time remains at least partly in force, as what a scientific theory elaborates, then the scientific theory cannot say why it is a theory of time rather than of 'tonk'. The radical issue here is what we require of understanding. It seems reasonable to require that understanding is achieved when one way of thinking about *x* is seen to be congruent with another equally compelling way of thinking about *x*; to require, in other words, 'the mind agreeing with itself'. On that requirement, a theory of (the unreality of) time recommending replacement does *not enhance understanding*. It may, indeed, be possible to make out a sense of 'true' in which one or more of these dislocated scientific theories is true: it might, then, be true that physical time is 'bent', or an eleven-dimensional continuum, or 'begins'. But it might also be true that truth-value will not survive for long the extinction of the value of truth to understanding.

The following two doctrines are ones to which all phenomenologists including Bergson would assent:

(i) fundamentally real time is experienced time;
(ii) fundamentally real time is not quantitative.

That (ii) is true is evident from a cursory consideration of our understanding of clock-time. A necessary condition of any temporal measurement is that two observations be made at different times, one later than the other. Unless we understand, without the aid of clocks, what it is for one clock-event to be later than another, we will not understand what it is for it to be five minutes later than another. In other words, we have an understanding of time prior to operational definitions of time (prior to clock-time).

(i) and (ii) neither entail nor exclude
(iii) quantitative time is non-fundamentally real

since (i) does not say that experienced time is *the only* real time (but just that it is *fundamentally* real). Heidegger, as I have said, espouses (iii), but compatibly with (i) and (ii). The quantitative time of Aristotle "has its own right"[13] and is not spatialized:

> The time 'in which' occurrent things arise and pass away is a genuine time-phenomenon and no externalization into space of a 'qualitative' time, contrary to what the inadequate and ontologically totally-undefined time-interpretation of Bergson would have us believe.[14]

In this chapter I want to construe Heidegger's attempt to reduce Aristotelian time to the original (fundamental) temporality of Dasein as a first move in a phenomenological elaboration of the concept of time from the bottom upwards. After some clearing of the ground (see pp. 170–174) the immediate horizon of this inquiry is set by the question (pp. 174–196) whether temporal processes can be quantified without falsification (spatialization). If they can, a phenomenological clarification of the evolution of the concept of time into and within science is in principle possible. The next horizon (pp. 196–199) is set by the status of the phenomenological claim that experienced time (Dasein's or other) is not just fundamental, but also real time. If it is, then the fact that time is, according to the phenomenologists, exclusively human will constitute no impediment to a realist interpretation of a science rooted in it.

The status to which phenomenology aspires *vis-à-vis* science may be brought out through a contrast with Kant's position on science. For Kant too time is exclusively human and science rooted in it (as well as in other exclusively human attributes). But because Kant further claims that this time is 'subjective' in a sense implying an *ontological deficiency*, he concludes that the science rooted in it could be true only of a merely phenomenal world, therefore closed to a realist interpretation. As against this, the phenomenologists take 'subjective' (a word they dislike) to mean strictly 'grounded in a subject' or otherwise *essentially connected with human being*, with no connotation of ontological deficiency; whence it follows that the 'subjectivity' of time is consistent with a realist interpretation of science.

Is time measurable?

Since time is a quantity in all branches of physics, the claim that physical time is real is as good as the claim that time apriori can be real (thinkable, non-contradictory) as a quantity. The first question that needs to be addressed is whether claiming that time can be a quantity (the 'quantifiability thesis') is the same thing as claiming that it is measurable.

To understand it to be a literal truth that time is measurable is to suppose that time-quantities apply to time itself, to the change *of* time or time-form, as distinct from the change of things or events *in* time. Let us call this 'the strong measurability thesis'. It is to be distinguished from 'the weak measurability thesis', which is just the claim that time-quantities logically can be true of physical processes – that these endure quantities of time. It is also to be distinguished from an even weaker, philosophically uninteresting, sense of 'Is time measurable?', arising from the question whether perfectly accurate (natural or man-made) clocks are available. What is being asked here is whether it is *technically* possible to assign *accurate* temporal quantities to natural processes assumed (by the weak measurability thesis) to be apriori measurable. Failure to distinguish between these three senses of 'Is time measurable?' has led to some talk at cross-purposes in the literature. For example, when F. Waismann remarks that "time would not be measurable if we happened to live in a world in which perfectly periodic processes were not available",[15] he appears to be suggesting that time's being or not being measurable is essentially a matter of luck. And so he would appear to be in disagreement with the view – represented by E. Harris and to which I shall come in a moment – that it is a matter of logic and not of bad luck that time is not measurable. Yet the two philosophers do not really disagree. The semblance of disagreement arises from Waismann choosing on this occasion to use the word 'time' to refer to the durations of physical processes assumed (by the weak measurability thesis) to be in principle measurable, while Harris uses the same word to refer to time itself, identified with a clock-motion. Both would agree that physical process are in principle measurable and that time as the clock-motion measuring them is not.

The technical interpretation of 'Is time measurable?' is by far the most prevalent, inasmuch as the scientific literature on time-measurement by far outweighs the philosophical. Under this interpretation the issues cluster adverbially around the verb *measure*: are physical processes exactly, absolutely, non-conventionally, commensurately measurable?[16] The philosophically basic question (answered negatively by Bergson) whether time-quantities logically can be true of physical processes is almost never raised.

Another factor making for confusion is that we often speak of rate-measurers or clocks as moving at speeds. This is a loose way of speaking which does not distinguish between a motion *qua* common-or-garden motion and a motion *qua* privileged clock-motion. The revolutions of the earth have been measured and so has the speed of light, yet both serve as clocks in physics. This does not mean that time (clock-time) has been measured, however, for when the revolution of the earth on its axis is measured, it ceases to be a motion in the role of universal measurer (to be 'time', operationally defined) and becomes one measurable non-clock-motion among others, all of them measured by whatever motion or process has been appointed to do duty as 'time'. When the revolution of the earth on its axis is measured, it is not measured *qua* time-standard, and when it officiates *qua* time-standard it is not measurable. ('Time' in physics flits about synchronically as well as diachronically.) Likewise, when an alarm clock is said to be 'going slow', it is no longer functioning as a clock-motion, but treated as a non-clock-motion whose rate is measured against a controlling clock assumed (without logical guarantee) to be in good working order and officiating as 'time'.

Turning now to Harris's arguments (which are standard among analytical philosophers), he asks rhetorically "How can we measure the pace of time?", and answers, "If we allege that time passes quickly, then we should be able to ask how quickly the last five minutes passed."[17] There are two claims here: (*x*) that time (identified here with clock-time: "five minutes") cannot be measured; and (*y*) that clock-time does not pass. I want to endorse (*x*) and contest (*y*). Harris's argument for (*x*) is as follows: "the fixed scale *is* time and we do not measure it any more than, when we measure a beam, it is the measuring rod that is being measured".[18] This relies on the general metrical axiom that measurement is always transitive, never reflexive: time measures something different from time, a measuring rod something different from the rod. So we have this argument for (*x*):

1 No measurer measures its own magnitude.
2 Time is a measurer (of magnitudes of change).
3 Therefore time does not measure time.

The argument is valid and the conclusion true.

The truth that (*x*) clock-time is not measurable does not prejudice the question whether clock-time passes. (*x*) entails that, if clock-time passes, it passes at no clock-rate; which means that there is *a* sense of 'how fast?' for which we can agree with Harris that it is absurd to ask how fast the last five minutes (part of clock-time) passed. But what is absurd here is not the question as such, but asking it *in the expectation of an answer in terms of clock-units*; for that there can be a non-absurd asking after 'how fast?' is shown by the fact that such answers as "very fast" or "faster than usually" are perfectly significant. This suggests some sort of dwelling of clock-time within lived time, to pursue which we need to look again at the standard argument against clock-time passing.

In considering the argument, I suppressed the premiss 'All change/passage has a measurable rate', which is strictly necessary to derive from the proposition that 'time moves' the (absurd) consequence that it must move in a superordinate time by which the rate of the time-motion would be fixed. This premiss seemed too transparent beneath 'Any time-motion must *qua* motion take place in an ulterior time' to justify the clutter of making it explicit to explain 'must'. However, my present purpose of defending the thesis that clock-time passes will be served by converting the standard argument in a way that puts the focus on rate. Thus:

(A) All change/passage has a measurable rate.
(B) Clock-time has no measurable rate.

Therefore:

(C) Clock-time does not change/pass.

Is (A) really true? It is false if either of the following pair is true:

(D) Clock-time passes, but at no rate.
(E) Clock-time passes, but at a non-measurable rate.

It seems to me that (D) is false; not because clock-time does not pass, but because it does pass at a rate (the non-measurable rate mentioned in (E)). So why, first, should we think that clock-time passes? Well, I do not see how it can be denied that, when the second-hand of a clock reaches the numeral 5, the point-events of its moving through the previous numerals are past. But given the identification of time with clock-time, this is *ex hypothesi* the fact of 'time itself' passing, not the passing of events *in* (measured by) time. The latter pass away, while clock-time does not. But the reason clock-time does not pass away, despite its losses, is not that it does not pass but that the motion with which it is identified is presumed to guarantee endless compensatory accretions, by never stopping; which was one reason why the ancients chose their clock-motions from among 'the motions of the eternal heavenly bodies'.

If this is right, and (D) yet false, then (D) is false because the second of the two constituent propositions, namely 'clock-time passes *at no rate*' is false. Why believe that clock-time passes at a rate *which is not measurable (x)?* Because clock-time is a standard against which the speeds of non-clock-motions are compared: how could it be a standard *rate* without being itself a rate (of sorts)? Why, if it is not a standard rate of change, is it not a standard of spatial magnitude? It must be a rate characterized by the admittedly mind-bending peculiarity of not being relativized to (measurable by) a standard, but a rate nonetheless.

But now it follows that (E) clock-time passes at a *non-measurable* rate is true, and that (A) is false. But with (E) we seem to be facing a monster: 'the absolute, *non-measurable rate* of passing clock-time'. All I want to do here is suggest the monster's affinities with Husserl's absolute time-consciousness; not so much by way of taming it as of showing that its ancestor is our lived time.

Without the aid of a clock I know that the dog I see chasing a horse is running slower than the horse. And my wife agrees. Are we, then, equipped with inner clocks? I think what has to be said is this: as absolute consciousness we possess, or rather are, an intersubjective time-standard for roughly assessing the relative rates of physical processes. 'Didn't time (a physical clock-motion) pass quickly today?', 'Didn't the last five minutes pass quickly?' To rule these questions out of order we have to caricature the inquirer as asking for a *clock*-measurement of just how much more quickly. But philosophers who do that are playing the wrong language game.

In the game in which the question is asked in everyday life, clock-time is not thought of as passing in clock-time, but rather in time-consciousness (which assesses its rate of passage as 'very fast', 'slower than usual', etc.). Given that clock-time is contained by time-consciousness, it follows that the latter "cannot be measured; there is no clock and no other chronometer for it" (ZB, 339). Clock-time measures neither clock-time nor time-consciousness. The containment of clock-time by time-consciousness leads to no regress, since time-consciousness does not duplicate clock-time by giving *clock*-measurements of it: the buck stops at "very fast", etc., which, when said of the passage of five minutes, is not a clock-measurement of a clock-measurement – not a clock-measurement at all.

It is noteworthy that Husserl never denies that time-consciousness is a sort of standard 'rate', a proto-clock. All he denies is that it can be speeded up or slowed down: it "runs [i.e. passes] just as it runs" and can run neither 'faster' nor 'slower' (ZB, 74). Is it not to this proto-clock of consciousness that we owe our concept of a clock-standard whose function is to enhance the accuracy and intersubjectivity of our assessments of empirical durations? If it is, we should expect clock-time to pass, and at a non-measurable rate; and to pose, as it does under scrutiny, the same intractable problems of description as time-consciousness.

Are time-quantities real?

The conclusion (*x*) that time is not measurable proves only that time-quantities, if they exist, do not apply to time (but at most to physical processes, events, etc.). I turn now to the question whether time-quantities do exist.

The question is not whether time inheres in the external world, but whether time-*quantities* are true of physical processes. That they are is the assumption on which almost all physical (time-)theory rests. Like the supposition that the laws of physics conform to a single time-scale, the quantifiability of time is not something that could be part of the informative content of a theory *in* physics. It is rather a pragmatic postulate of physics getting off the ground. The quantifiability thesis may therefore be approached philosophically without regard to the content of particular physical theories of time. The issue remains the same whether it is set in the context of Newtonian physics, relativity theory, thermodynamics, quantum mechanics etc., since these are all time-quantifying contexts.

For time-quantities: the Kantian paradigm

Although the explicit formulation of the quantitative conception of time goes back at least as far as Plato, its most impressive (because 'deductive') vindication is to be found in the link between time and quantity forged by Kant in his deep, difficult and ingenious Schematism of the Pure Concepts of the Understanding. Having shown that concrete experience involves the subsumption of sensible intuitions under the pure categories of the understanding (quantity, quality, unity etc.), Kant raises the question of the conditions of possibility of that subsumption. As well he might, considering that 'intellectual' categories seem disinclined to cuddle up to sensible intuitions. The former must somehow fit the latter as a glove fits a hand: "in all subsumptions of an object under a concept, the representation of the object must be homogeneous with the concept" in the way, for example, that "the empirical concept of a *plate* is homogeneous with the pure geometrical concept of a *circle*".[19] There must be some "third thing",[20] neither pure understanding nor pure sensibility, to mediate the subsumption of the latter under the former. And this third thing, Kant reckons, is time. On the one hand, time is homogeneous with sensible intuition in that it is the non-sensible (apriori) form of sensible intuition. On the other hand, time is also homogeneous with the aprioricity of the categories in that "a transcendental determination of time . . . is universal and rests upon an apriori rule".[21] For example, assuming with Kant that substances are permanent, then our understanding of the category of substance connects with intuition inasmuch as it is a property of formally intuitive time not to be transitory. The category of substance, therefore, is schematically anticipated in the non-transitoriness of time. Likewise, the categories of actuality and necessity are schematically anticipated in, respectively, the possibility of an object existing at a determinate time and at all times. Time is the raw stuff from which the pure apriori (as distinct from reproductive) imagination works up unpicturable schemata in one-to-one correspondence with categories. Time is the soft intuitive underbelly of categories. Now, I think it is palpable that, if we suppose time to be a sum of parts put together by a subject's act of synthesis, then we might expect that this synthesis of parts too is a property of time fitted to serve as the schema of some category. And indeed Kant finds in number, *as produced by the synthesis of time parts*, the schema of the category of quantity:

> the pure *schema* of magnitude (*quantitas*) as a concept of the understanding, is *number* understood as what has been generated by the subject's activity of synthesizing time-units. Number is a representation which comprises the successive addition of homogeneous units . . . therefore simply the unity of the synthesis of the manifold of a homogeneous intuition in general, a unity due to my generating time itself in the apprehension of the intuition.[22]

But notice further that this numerical time is generated with essential reference to the synthesis of the manifold: as something transitive, not as something that could apply to itself: "[t]he schema of magnitude is the generation (synthesis) of time itself in the successive apprehension of [the parts of] *an object*".[23]

Questioning time-quantities: Augustine

Anyone familiar with the relevant section of Augustine's *Confessions* will know that he repeatedly asks himself whether 'time is measurable'. Yet here I shall be considering his remarks under the heading of whether time-quantities are real (thus capable of being true of physical processes). This is another brush with the verbal awkwardness we encountered before, apropos of Waismann and Harris. There is ample evidence that by 'Is time measurable?', Augustine means 'Can there be time-quantities true of physical movements?' He ponders whether we can measure the duration given *to syllables* in reciting verse;[24] and he says clearly that "it is *by time* that I measure [not time but] the duration of the movement" (C, 238; emphasis added). Despite his talk of 'measuring time', Augustine is not wondering whether time measures its rate of passage (though we shall have to bear with this talk in what follows).

Augustine defends a proposition that reveals him as a phenomenologist *avant la lettre*: that time has its source in the human soul and is not to be identified with, or made to depend upon, motion: time is not a periodic motion such as that of the 'celestial bodies' – the scientific clock-time of his day. "Let no one tell me then that time is the movement of the heavenly bodies" (C, 238). "I have heard a learned person say that the movements of the sun, moon, and stars in themselves constitute time. But I could not agree" (C, 237). To be in time a body has no need to be in motion: we measure the duration of a body's rest just as we measure the duration of its movement and rate (assuming time-quantities to be real):

> a body may at one point be moving, at another point at rest. We measure by time and say 'It was standing still for the same time as it was in movement' . . . or any other measurement we make, whether by precise observation or by a rough estimate (we customarily say 'more or less'). Therefore time is not the movement of the body.
>
> (C, 239)

This argument seems unanswerable as long as rest is defined in some phenomenological way. In the hope of rescuing the Aristotelian doctrine of the dependency of time on motion, we might try defining rest non-phenomenologically as the state of a body travelling at infinite velocity. Such a body would be 'at rest' in that it would never be *en route* between a spatial point reached and another not yet reached: for any point you care to designate, it's always already there. But is this not just a convenient sort of 'rest'?

A body is in *our* time, which is that by which its motion or rest is measured. As I have suggested, this subjective measuring (as 'more or less') banally takes place in the absence of clocks: "if the heavenly bodies were to cease and a potter's wheel were revolving, would there be no time by which we could measure its gyrations and say that its revolutions were equal?" (C, 237). Of course there would still be time, for "a body's movement is one thing, the period by which we measure it another. It is self-evident which of these deserves to be called time" (C, 239). What deserves to be called time is the time of our consciousness (of the movement of the body).

For as long as he is concerned to free time from motion and ground it in consciousness, Augustine gives a free ride to the assumption that time-quantities are real. When he confronts this assumption, however, he is beset with doubts. The duration of a sound cannot be measured before it begins, for then it has no duration. Nor can it be measured after the sound has ended, for then the sound exists no more. Nor can it be measured in the now when it is going on, for "the present occupies no length of time" (C, 236):

> In what extension do we measure time as it is passing? Is it in the future out of which it comes to pass? No, for we do not measure what does not yet exist. Is it in the present through which it passes? No, for we cannot

> measure that which has no extension. Is it in the past into which it is moving? No, for we cannot measure what does not exist.
>
> (C, 236)

Yet Augustine is not prepared to conclude without qualification

> that time quantities are unreal: "nevertheless we do measure periods of time . . . Yet we measure periods of time".
>
> (C, 241)

Something has to have extension if it is to be measured, but in the case where the extension would be temporal it cannot exist because only the extensionless now exists. Augustine surmises that only the mind escapes this paradox: "it is not the syllables I am measuring but something in my memory which stays fixed there. So it is in my mind that I measure periods of time" (C, 242). The 'time' Augustine is certain we do not 'measure' is "time as it is passing" (C, 235), which has no extension; and the time he thinks we can 'measure' is the extended representation in memory of extensionless passing time: passing time is never 'there', whereas remembered time is passing time re-presented in extension. But time as it is passing is a more fundamental manifestation of time than remembered time. Does Augustine's position not amount to an admission that only the shadow-show of time is measureable? When he says: "I have come to think that time is simply a distension" (of the soul) (C, 240), he implies that past time deserves to be called 'time'. Elsewhere, and in conclusion, he is not so sure: "therefore either this [remembered time] is what time is, or time is not what I am measuring" (C, 242).

Time-quantities denied outright: Bergson's argument

While Bergson's argument shares a common core with Augustine's, it brings in considerations of counting and spatialization in a way that non-suits any complaint that he simply transposes Augustine's argument from a sceptical into a demonstrative key:

> [There can be] no succession which results in a sum. For although a sum is obtained by successive focusing on different terms, it is still necessary that each of these terms remains when we pass on to the next one – waits, so to speak, for us to add it to the next one; and how can it wait if it is only an instant of time? And where would it wait if we do not locate it in space?[25]

Notice that Bergson has tacitly assumed that the now has no extension. Augustine reasoned on that premise too, but he put it up front. We shall see Heidegger challenge it.

The core of Bergson's argument is simple. Time quantities could not come ready-made, they exist only if moments can be counted up. To be counted (as distinct from merely enumerated), moments would have to be present together before the mind. But they are not, owing to their evanescence. So they cannot be counted, whence it follows that there can be no extensive magnitudes of time.

What we count under the illusion that we are counting the moments are their "symbols" (DI, 56), the moments projected into a mental space which secures them against evanescence:

> when we add to the present instant those which have preceded it, as happens when we add units, it is not on those units themselves that we operate, since they have vanished for ever, but the lasting trace which they appear to us to have left in space as they pass through it.
>
> (DI, 59)

Despite agreeing with Augustine on the impossibility of passing time forming a temporal extension, Bergson would dismiss as an illicit spatialization the distension in memory which Augustine thinks might do duty as a non-evanescent time-extension.

The scene for the illusion that there exist time-quantities is set, according to Bergson, by assimilating time to physical motion: not to pure motion as an indivisible and therefore time-like process, but rather motion as "the space moved through". Dividing up that space gives us a Bergsonian 'mixture', namely spatial points in temporal drag – the instants of mathematical time. Why not simply spatial points? Because the temporal origin of these hybrid points must linger on in the mind, since otherwise "the point would be only a point, there would be no instant";[26] "it is easy to see why the dimension of space which has come to replace time is still called time . . . Our mind, interpreting mathematical time, retraces the path it has travelled to obtain it."[27] The instants of mathematical time are thus spatial points masquerading as temporal points *in memoriam* of the pure time from which they are derived.

The location of countable items in space is necessary, Bergson thinks, to ensure not just their simultaneous presence before the mind, but also that they are distinct in the non-qualitative way he supposes to be required for counting. Countables must be both distinct and synthesizable; which means they cannot be distinct by quality, since no synthesis of qualities gives a quantity: when I count the animals in the farmyard I am not producing a synthesis of henhood, horsehood, doghood, etc.; whence Bergson concludes that to enter the count each animal must be stripped down to a bare *it*, all of them treated as "absolutely similar to each other" (DI, 57). How, then, are these homogeneous items distinguished from each other as they must be? Bergson's answer is that they are distinguished by the positions they occupy in geometrical space. But it is as identified with spatial points that they are also perfectly alike, and perfectly simultaneous. Are points in space not perfectly alike while being perfectly distinct and also perfectly simultaneous? So all the conditions of counting might seem to point in a mutually confirming way to space being the milieu in which whatever things are countable must be placed.

Of Bergson's two reasons for locating countables in space – the need for them to be distinctly homogeneous and to co-exist simultaneously – the latter is by far the more compelling. For the notions (i) that only things that are perfectly alike can be counted together; and (ii) that spatial points are perfectly alike were, both of them, superannuated by Frege shortly before Bergson embarked on his philosophical career. As to (i): while it is true that counting essentially involves collecting units on a basis of sameness, in this sameness does not consist in the units being perfectly alike, it consists in their being seen as falling under some single concept. The function of qualities, or of whatever non-qualitative determinations distinguish things, is to specify *cases* of a concept; which cases (instances of the concept's existential import) are what gets counted. Not things as qualities, nor things stripped of their qualitative differences, but things as cases specified by their qualities (typically). How similar or dissimilar the embodiments of the cases may be is therefore irrelevant, for it is senseless to speak of existential instantiations resembling or not resembling one another. The qualities of things do not get counted, for it is true that we cannot by counting make a *qualitative* unity out of a white and a black and a ginger cat. But nor is it totally alike units obtained by abstraction from things having qualities that get counted: "the white [cat] still remains white just the same, and the black one black".[28] As to (ii): the very notion of two things being perfectly alike offends against Leibniz's principle of the identity of indiscernibles (if everything true of A is also true of B, then A is identical with B and there are not two things). Accepting this principle, we might deduce the impossibility of perfectly similar units being in space from the impossibility of their existing at all. More empirically, however, we might point out that, however matters may stand in other domains, it is demonstrable that spatial points are only dissimilar, not perfectly identical as well. That they do not all stand in the same relations – B differs from A in that B is next to C whereas A is not – Bergson would admit, since he describes space as "a principle of differentiation other than that of qualitative differentiation" (DI, 71). How, then, can he claim that spatial points are also perfectly similar? It seems to me that the answer is this: just because spatial points have no positive (qualitative) characterization independently of the relations in which they stand, they appear identical with each other once they have been abstracted from these relations, as Frege pointed out: "it is only once considered in themselves, and neglecting their spatial relations, that points of space are identical with each other".[29] But then they are no longer points of space at all. So to conclude: there is no need for countables to be perfectly alike and there is no chance of them being such when symbolized by spatial points.

Bergson husks over the working Augustinian core of his anti-quantifiability argument with an idle theory of number which drew Russell's withering remark that "Bergson does not know that number is".[30] All Bergson needs to prove is that *counting* rests on the juxtaposition of simultaneous items in physical or mental space; he does not need to identify *number* with the collections produced by counting them: what his argument requires is that we count in space, not that number be tied to space (i.e. definable as a collection). Yet he defines number as "a collection of units" (DI, 56), says that the number 3 is the collection of 1 + 1 + 1 (DI, 61) and that "space is the matter with which the mind constructs number" (DI, 63). It does not seem to occur to Bergson that *a* spatial position is an item presupposing number. Russell retorts that his definition of number confuses abstract number (what particular numbers have in common), particular numbers (3, 4, etc.), and the collections (such as the 12 Apostles or Bergson's arrays of spatial points) in which particular numbers are exemplified. Far from applying to number in general, his definition does not even apply to particular numbers, but only to collections. A collection, which is produced by counting, is not identical with the particular number it exemplifies, far less with number in general. That number cannot be identified, as by Bergson's definition, with the result of counting is evident from the simple consideration that two collections, each of three different kinds of thing, do not give us two kinds of three. It is admittedly arguable that we would not have the abstract concept of three if we did not have the concept of a threefold group. Even so, the former could not be defined in terms of the latter.

But to come back to the alleged necessity for countables to exist simultaneously, one response to this has been to invoke stipulative devices in the spirit of Wittgensteinian ruling. It has been claimed that the difficulty rests on an unnecessary analogy between object-wholes, such as a house, and temporal wholes such as an event taking time to happen. A house is measurable because it exists as a whole whose parts are present together. But it is questioned whether this obliges us to make it a requirement of assigning time-quantities to events that they too must be wholes whose parts are present together. The proposal is that we might rule that events are a different sort of whole, one whose parts need not be together. Alternatively, if we find ourselves unconquerably opposed to the notion of a whole whose parts are not together, might we not rule instead that an event has a temporal magnitude despite being no kind of whole? The advantage of these rulings is supposed to be that, on either, events come out with temporal magnitudes. But notice that we cannot *think* the stipulated situations. To ease this cramp it is suggested that "it is not desirable that we should make a fetish out of intelligibility or consistency".[31] But there the cat is out of the bag: the problem is to be got rid of, not by solving it (for any solution would be in terms of intelligibility) nor even by dissolving it (for no linguistic muddle has been cleared up), but by giving up philosophy. What is proposed is a cure for a headache by cutting off the head.

To assess the Augustine–Bergson core argument we need to go back to its atomistic premise, which was that moments cannot be together in time before the mind. But then it follows that there can exist no genuine time-order, since this involves apprehending moments in relation to each other: time-experience must consist in a multiplicity (not even that) of mutually oblivious moments. In response to this implausible result, it should be pointed out that our conviction that a time-order exists is far stronger than Bergson's argument against it, to wit:

> we cannot establish an *order* between terms without first of all distinguishing them, without then comparing the places they occupy; and so we perceive them as multiple simultaneous and distinct; in a word, we juxtapose them, and if we establish an order in the successive we have in fact turned succession into simultaneity and projected it into space.
>
> (DI, 76)

So watching six milk bottles falling one by one (in temporal order) off a wall would be an experience ontologically on all fours with seeing them sitting on the wall. Bergson's denial of temporal order does violence to contrasts that are too palpably basic to mental life to be denied without gross implausibility.

We saw that one of Bergson's reasons for wedding counting to space was that space provides a non-qualitative principle of differentiation. Another is that he thinks, rightly, that countables must be simultaneously contemplatable in a single act of mind; whence he concludes, wrongly, that only space could meet this requirement:

> If I represent these sixty swings [of a pendulum] to myself all at once and by a single act of mind, I exclude *ex hypothesi* the idea of succession: I think, not of sixty beats succeeding one another, but of sixty points on a fixed line.
>
> (DI, 78)

Surely not *ex hypothesi*. All sorts of relations are compatible with terms being represented simultaneously. It is true that if I represent two things as simultaneous, I represent them simultaneously. But it is not conversely true that representing two things simultaneously entails representing them as simultaneous.[32] When I compare the game of golf I played last week with the game I played this week, I simultaneously represent the two games as successive. Thus, although I am simultaneously representing them, I am not representing them in space, but in time. (Were the representation spatial it would not contrast, as it obviously does, with the case of conjuring up a mental map of the relation between two towns – Bergson's thesis obliterates inalienable intuitions of contrast.) All the possibility of counting requires is (i) that the now-origin of events replace spatial position as a differentiating principle; and (ii) that there can be some genuinely temporal sense in which events thus differentiated can be together before the mind. Now this prospect leads us to Heidegger.

For and against quantifiability: Heidegger's arguments

The context of the argument 'for' is Heidegger's interpretation[33] of Aristotle's canonical definition of time as "the number of motion with respect to earlier and later". The foci of interest will be Heidegger's reading of the nature of the Aristotelian now and his claim that Aristotle's numerical time is grounded in the original ecstatic temporality of Dasein. The German philosopher's "free discussion" (as he acknowledges) does not square with everything relevant in the *Physics* and *Metaphysics*, and some of his exegetical claims seem to turn on fine points of Greek philological scholarship[34] on which I am not competent to comment. So I shall proceed on the understanding that we are dealing with *Heidegger's* conception of now and its ground, vehicled through his interpretation of Aristotle as sharing it.

The dimensional now of (Aristotle's) common time

The motion mentioned in Aristotle's definition is a special case of change (*metabolē*) as such: "The most general character of motion is *metabolē*, that is, change [*Umschlag*], or better, transition from something to something" (GP, 343). According to Heidegger, this structure 'from something to something' is abstractly temporal and defines change in general, not just physical motion, though the latter is Aristotle's preferred example (GP, 343). It is Heidegger's contention that Bergson's charge against Aristotle of having spatialized time resulted from his misinterpreting Aristotle's dimensional structure 'from something to something' as modelled on physical motion, construed as corresponding with the continuity of space:

> Bergson . . . says that time, as understood by Aristotle, is space. He was misled into this inappropriate interpretation as a result of conceiving [Aristotle's] continuity in the narrow sense of the extensive magnitude of space.
>
> (GP, 328)

> especially with Bergson the Aristotelian concept of time has been misunderstood, in so far as he from the outset construed this dimensional character of time as spatial extension.
>
> (GP, 343–344)

The argument, though not spelt out by Heidegger, seems to run as follows. Bergson misreads Aristotle's 'from something to something structure' as applying to physical motion repeating the continuity of space. He is then able to argue plausibly that Aristotle reduced the 'from . . . to' to a sum of spatial points. But he could not plausibly have maintained this had he appreciated that Aristotle's 'from . . . to' applies no less to the qualitative change of an apple from green to red, or of a man from being happy to being sad. Since neither of these changes involves change of spatial position at all, the scope for arguing that Aristotle reduced time to 'the number of' a motion in space is foreclosed.

The fact about now which particularly interests Heidegger is that it is, as Aristotle says, "the link of time". Or as Heidegger echoes with approval:

> the now . . . has the character of a transition. The now as such is already a crossing [*das Übergehende*]. It is not a point next to another point, for which two points a mediation would be needed, but rather it is in itself already the transition.
>
> (GP, 352)

This transition Heidegger variously dubs the "spannedness", "stretchedness", "extensivity" and, I think best, "dimensionality" of now. The 'now' is originally multiple, inasmuch as it is originally the transition between now-no-more and not-yet-now: the 'extension' of time is not the result of the addition of durationless now-points.

To affirm the dimensionality of now is to deny that it is a time-point in the obviously silly sense in which time-points are thought of as indivisible "bits" separated by voids of non-time. But what of the more sophisticated conception of the point as a beginning and end? Heidegger does say that "the now is a beginning and an end" (GP, 354); and he acknowledges that:

> the assimilation of the multiplicity of nows – nows taken as transition – to a multiplicity of points (line) has a certain validity only when we take the points of the line as forming a beginning and an end, that is, as constituting the transition of the continuum and not as juxtaposed bits available for themselves.
>
> (GP, 352)

The assimilation is grounded in the fact that external processes don't just start and stop, they start and stop now: with the coming to an end of a process, such as motion, a limit comes to be coincidentally associated with a now, a coincidence which 'marks' now as an end. But that extraneous determination does not detract from the essential limitlessness of now:

> The now – and that means time itself – is, says Aristotle, *by essence never a limit*, because as transition and as dimension it is open on one side towards the Not-yet and on the other towards the No-more. It is only accidentally, with reference to something which stops *in* a now and at a definite time-point, that the now is a limit in the sense of a termination,

> of a completion, of a no-further. It is not the now as now that stops, but rather the now is essentially not-yet, already as a dimension related to what-is-to-come; whereas it is quite true that a movement determined by saying 'now' can stop in that now. With the help of now I can mark a limit; the now as such, however, has no limit-character, inasmuch as it is taken within the continuum of time itself.
>
> (GP, 352–353; see also GP, 355)

Now is always different in so far as it is adventitiously individuated by the stopping or starting that *happens* coincidentally with it. But it is also, and essentially, always the same, in that it is not limited by these individuating stoppings and startings. Now does not limit, but is adventitiously limited: essentially 'it is always the same old now' (see above, pp. 149–150). Now is the same in respect of its 'whatness' (*essentia*, *Wassein*), different in respect of its 'how-being' (*existentia*, *Wiesein*) (GP, 349–351). As counted off a motion, now is different with every correlated place: "because of the transition of the thing that moves, the now is always other, that is, a progression from one place to another" (GP, 350)."Because the now is what is counted in the transition, it is itself always other along with the thing that moves" (GP, 354). It is this now in its externally-determined (accidental) how-being, as always-different, that is counted in the perceptual following-through of a motion.

The non-arbitrariness of associating the essence of now with sameness (with transition) is well brought out by the paradoxical result of A.J. Ayer's attempt to define it in terms of its more palpable 'empirical' how-being, as "the class of all the events which are contemporaneous with *this*, where *this* is any event we choose to indicate at a given moment".[35] As Richard Gale rightly pointed out,[36] on this definition we should have to go on calling this class of simultaneous events 'now' long after they are past in the ordinary tensed sense.

Because now is not limited (but stretched), it spans the whole of time: now is not part of time. We have already seen Heidegger slip in the statement that "the now . . . means the whole of time". Or more explicitly:

> [Time] is, as now, not *part* of time, such that this time would be composed of now-parts, but rather each part has the character of transition, that is to say, it is not actually a part . . . the now is thus no part of time but always time itself.
>
> (GP, 344)

The stretchedness of now shows up in its familiar elasticity: depending on my present interests, this second can be now, also this day, this month, nowadays: "This difference in the extent of the dimension is only possible because now is in itself dimensional" (GP, 352). Now as this second is no differently now than now as nowadays, in something like the sense in which a 5-foot section of a pole is no differently a section of its *vertical dimension* than a 10-foot section, despite the latter being a greater section *of the pole.*

Aristotelian time (time as inseparably connected with the perception of change) is time generated by counting (cf. Kant's version of the generation of time together with number). Time is "what-gets-counted-up off motion" (*das Gezählte an der Bewegung*): a row of books exists whether we count them or not, whereas Aristotle's time exists only as counted-up off a motion or other change. However, this time "in the sense of the counted nows can itself count in the sense of measuring" (GP, 354). That is, a counted motion can serve as a clock for counting other motions. Clock-time is therefore 'what-is-counted-up in turn counting' (*das Gezählte-Zählende*: GP, 348, 353, 354). Correspondingly, Heidegger speaks of "number" both as *numbered* number, specifically as the number *of* counted nows ("time is number in the sense of counted nows", GP, 354), and as *numbering*, specifically as the clock-numbering of motions other than the clock-motion. But the crucial notion in all this is that "time is *number because it is transition*" (GP, 354; emphasis added). I think a clue to the meaning of this gnomic statement is given by Heidegger's deliberately unidiomatic practice of omitting (as above) the indefinite article before 'transition' (*Übergang*). Now (time) is originally multiple, yet has no particular number (corresponds neither to this second nor to nowadays) until somebody counts (quantifies) the 'hows' of now. As such, now would be *potential* numberedness, transitionality that is not in itself *a* transition or *many* (now as either would be limited). Now (time) is a dimension *for* the 'more' and less', not something which is in its essence determinately more or less. When Heidegger says (as above) "the now is what gets counted in the transition [i.e. not the transition]", he is, I think, speaking exactly. What gets counted are the 'hows' of now, presented in a dimension of betweenness that brings them together *in time* before the mind.

Because now is essentially dimensional, the counting the tokens of now-here (the 'hows' of now in the case of a motion) does not fragment the unity of the motion presented in the now-dimension. With every adding of a 'now-here' to a previous total, we encompass the total state of the motion:

> Now is thus no part of time but also time itself, and because it is no part of time, a movement, in so far as it is measured by time, is also not fragmented. Because now is transition it is capable of making accessible the motion *as motion*, that is, in its continuous transitional character . . . [o]n the basis of this transitional character [the now] has the peculiarity that it measures the motion as such, as *metabolē*.
>
> (GP, 354–355, 360–361)

Whence it emerges, that the Augustine–Bergson core argument rests on the false premiss that now is dimensionless. On that premiss a motion can have no (measurable) extension because the time-form in which it is represented has none.

Thus far, nothing has been said of "with respect to earlier and later", which complements "time is the number of motion" in Aristotle's definition. The *earlier* vs *later* contrast is conceptually and linguistically distinguished from the 'before vs after' contrast, at least in ancient Greek, German and English. Aristotle himself says that "the distinction of 'before' and 'after', then, holds *primarily* . . . in place" (*Physics*, IV, 219a; emphasis added), while Heidegger reports that "for Aristotle before and after have *non-temporal* meaning", since "they apply to the succession of places" (GP, 349; emphasis added). In this he goes too far, for Aristotle immediately adds that "also in time the distinction of 'before' and 'after' must hold, for time and motion must always correspond with each other" (*Physics*, IV, 219a): correspond in the sense that a moving body could not be at a place 'spatially' *before* another if it were at the first temporally *later* than the second. But let that pass. The important points are, first, that the places apprehended through observing a motion are temporally articulated in a way places considered apart from motion are not: we do not grasp a sheer juxtaposition of There and Here:

> but rather the from-There-hither is something *before* and the from-Here-thither is likewise not any Here, but, being a Here-hither for the next Here, is something *after*. When in this way we see the multiplicity of places within the horizon of from-There-hither-Here-thither and run through within this horizon the individual places, while seeing the movement, the transition, then we *retain* the place first run through as the *From-There-hither* and we *anticipate* the next place as the *There-thither*. Retaining the before [*Vorige*] and anticipating the after [*Nacherige*] we see the transition as such.
>
> (GP, 347)

The space being moved through is, we might say, 'smudged', quasi-temporalized, to fit the transitional time-form of motion.

Second, Heidegger understands the succession of places ('from *before* to *after*') to be less purely temporal than the fundamentally-temporal retaining-anticipating (*earlier* vs *later*) structure underlying it:

> The 'with respect to before [*vor*] and after [*nach*]' and the 'within the horizon of earlier [*früher*] and later [*später*]' do not coincide: the latter is the interpretation of the former . . . the experience of before and after in itself presupposes in a certain way the time-experience, the earlier and later.
>
> (GP, 349)

We have now identified two horizons within which the perception of time as what-is-counted-up off motion is imbricated: the *spatio*-temporal before-after or 'hither-whither' horizon, and the *purely* temporal 'from earlier to later' horizon within which the first is contained. Inasmuch as the counted number of now-place correlations falls within that first horizon, what is counted is not static spatial points, but a motion *qua* motion. Inasmuch as that first horizon itself falls with the second, purely temporal, horizon, the spatio-temporality of motion is reduced to – 'interpreted into' – a fundamental space-free (autonomous) time. This is the early Heideggerian doctrine of the reducibility of space to time, upon which rests his thesis that quantified space-*linked* (i.e. Aristotelian) time is not spatialized time. Or more specifically: the doctrine neutralizes the danger that in the correlation here–now, the now might be spatialized by the here.

The train of thought here is brought into sharper focus by Heidegger's ingenious interpretation of the apparent circularity of Aristotle's definition, which defines time in terms of the temporal words 'earlier' and 'later'. "Is it worth bothering with a definition which bears the stigma of the grossest of logical mistakes?", asks Heidegger. Yes, it is, he answers, once we see that Aristotle is not offering an "academic definition" but an "*access-definition*" (GP, 362), a method for finding out what time is. To see this, suppose Augustine had addressed to Aristotle his famous question, "What then is time?" According to Heidegger, Aristotle would have replied roughly as follows: 'Observe a motion, with a view to counting the places it traverses. With your counting of places you will find yourself also counting something co-glimpsed through your seeing the body traverse places. That something is "now", time. Through my method you are brought to "see" time, you gain access to time as what-is-counted-up off motions'. As Heidegger put it: "the counting perception of motion as motion is at the same time the perception of what-is-counted-up as time" (GP, 362).

But it is also and at the some time the espying of a time-horizon pre-understood, as a different and deeper sort of time, to the counted or common time referred to by the subject-term of the definition. This deeper, horizonally co-apprehended time, is the time referred to by "earlier and later", whose difference from the counted time remains invisible to a purely linguistic analysis of the definition. Once seen, the difference of course rescues the definition from circularity:

> admittedly earlier and later are temporal phenomena. But it remains to be asked whether what is meant by them is identical with what is meant by the subject-term of the defining proposition: time is time. Perhaps the second word 'time' says something different from, and more original than, what Aristotle himself means with his definition of time. Perhaps the Aristotelian definition of time is no tautology, but betrays simply the internal connection of the Aristotelian time-phenomenon – that is, of the common concept of time – with the original time we call temporality. (GP, 341)

No perhaps about it, really. In Heidegger's view, the charm of the definition lies in the fact that "in [it] that which he takes as time is interpreted out of time" (GP, 342). "The definition of time which Aristotle gives is so ingenious that it also fixes the horizon within which we encounter, with what is counted off the motion, nothing other than time" (GP, 240). The thought here is that, just as the observation of a motion 'gives access to' counted (common) time, so counted time in turn gives access to (is the horizon from which is 'sighted') the original pre-spatial time out of which counted time shows up, as out of its foundedness in the fundamental temporality of Dasein.

In order to get a fuller picture of the space-linked common now to be reduced, we need to consider some more of its characteristics, all of them congruent with its dimensional character. The question, phenomenologically, is, 'How does now originally show up for us?' What am I attending to when in the course of a busy day I look at my watch and say 'It is (now) 4 p.m.'? Heidegger's view of the situation may be represented by saying that he regards 'It is (now) 4 p.m.' as a shorthand for 'Now that it is 4 p.m. it is time to do such and such.' We are focused on what needs to be done (now), or on whether it is the right time to do something, or on whether we have time. We are reckoning concernfully with time. We are not focused on time as such (for time is not in the inspected watch). In other words, this now with which we are pragmatically concerned in everyday life is not an inspectible entity or event: "[w]hen we say 'now' we are never attending to the now as something occurrent [*Vorhandenes*]" (GP, 365).

Our everyday (experientially-first but not ontologically-first) relation to time is this concernful reckoning with it in a sense broader than counting. We reckoned with time ('cared for time') in this broad sense, which includes making pre-numerical estimations of magnitude ('we waited for ages'), long before doing so led to the invention of clocks as a means of standardization of specifically calculative reckoning (GP, 365, 368).

Understood as 'time to...', etc., now possesses a property which Heidegger calls "referential significance" (*Bedeutsamkeit*), which alludes to the essential pragmatic involvement of now in a world experienced as "a totality of relations ... possessing the character of in-order-to" (GP, 370); in a world structured as an equipmental complex serving Dasein's tasks. Referential significance is reflected in the fact that 'Now it is time to...' is followed by a verb. In acknowledgement of its worldliness, Heidegger dubs the dimensional now "world-time" (GP, 370), not to be confused with the natural time he rejects: "the world is nothing occurrent ... there exists no natural time" (GP, 370).

The now of world-time (the dimensional now) is essentially datable by an event. A sentence like 'Now that it is getting dark, it is time to go home' registers both now's datability by an event (getting dark) and its referential significance (go home). Likewise with future and past nows. When I say 'then' referring to a future time (then as *dann*), I always mean 'at a future now when...' (*dann wann*); and when I say 'then' referring to a past time (then as *damals*), I always mean 'at a past now when...' (English is embarrassingly deficient in temporal adverbs compared to German).

Essentially now is dat*able* and only accidentally is it dat*ed*. Were it essentially dated, it would be tied to a date, inelastic, an occurrent entity that comes and goes with the dating event to which it is tied, not the whole of time.

A fourth property of the common now of world-time is its publicness – the fact that it is the same now for all of us, irrespective of the variety of clocks by which it is dated. The now dated 6 p.m. in the UK is the same now as that dated 1 p.m. in New York. The publicness of now is nearly the same thing as its encompassing character (*Umhalt*). We think of objects as being outside of us and of now as outside of even objects, encompassing the totality of what-is (occurrent being). To see that the now is outside of everything is to concede that it is essentially no kind of occurrent thing or event.

The reduction of common time to Dasein's original time

The animal to be reduced, then, is this "full now phenomenon" (SZ, 424), the now as originally adorned with the considered properties, as distinct from the "naked now" (GP, 365), of whose nakedness more anon. However, a reasonably close kinship between what is to be grounded and its ground is a minimal condition of the possibility of any grounding project. To get a sense of the requisite kinship in the present case will be to glimpse how the full now-phenomenon (world-time) is already anticipated in (has analogues in) the present past and future belonging to the original temporality of Dasein.

By 'the present' of original temporality we are to understand the fact that Dasein "sojourns amidst occurrent entities" (GP, 376). This is a way of saying without invoking intentionality that human beings are 'outside of themselves', originally removed (*entrückt*) from themselves towards things (GP, 377), involved in the "ecstasy of the present" (*Gegenwart*). Original temporality is the time of what Heidegger calls "transcendence", in virtue of which Dasein finds itself in-the-world-alongside-things before (in no sequential sense, thus 'originally') intentionality aims at them.

The present temporalizes itself in immediate unity with the ecstasy of the original future and the ecstasy of the original past. The 'whither' (*Wohinein*) of each ecstasy is its horizon (GP, 378), the "openness" opened up by the ecstasy. The temporality of Dasein is therefore "ecstatic-horizonal".

To say that these ecstasies temporalize themselves in immediate unity with each other is to say that the whole of temporality temporalizes itself in one multi-dimensional 'explosion', so to speak. Though having different temporal values, the ecstasies are 'together' ("equiprimordial": SZ, 337; cf. Husserl's all-at-onceness), contrasting in this with tenses, which do not apply to an event all at once (if time is real). The equiprimordiality or co-originarity of the ecstasies means that this fundamental time is not a succession: "[t]emporalization signifies no 'succession' of ecstasies. The future is *not later* than the already-dimension [*Gewesenheit*] and the latter is *not earlier* than the present" (SZ, 350).

Dasein's present is its *making-present* (*gegenwärtigen*) the world towards which it transcends. This present is not present in the tensed sense exclusive of past and future. The present (*Gegenwart*) is not a punctual sensory now, but rather the presenting of the world within the horizons of original pastness and original futurity. Something that is retained or anticipated is present in Heidegger's sense inasmuch as it 'comes up' for Dasein as *being* an object of its concern, as not being nothing. Thus a dead friend I am thinking about is by my thoughts made 'present' because my thoughts are, as originally-temporal, making-present, in this case retentively-present-making (*behaltend-gegenwärtigend*).

The upshot of these summary considerations is that the original present is ubiquitous throughout time (appropriately, seeing that it grounds a now that is no part of time). It is not that part of time which *was* future and will *become* past (this original temporality does not pass). The original present always-already *has* a future which does not become present. And the original past is a past which was not a previous present.

The original future is not a property of an event or thing, but a relation of Dasein to *its* future being-in-the-world-alongside-things. In every anticipation (*Gewärtigung*) of a happening Dasein concomitantly projects itself towards its future self as towards its "ability-to-be-a-whole" (*Ganzseinkönnen*): "in the anticipation of the process itself, our own Dasein is always co-anticipated" (GP, 375). The idea here is that the original future as a mode of making-present is "the presupposition for the common concept of the future as not yet now" (GP, 375). It is things and events that are not yet now, not the original future. The original future (Dasein's *self*-anticipation) is the condition for the possibility of Dasein holding out expectations (*Erwartungen*) as to how things or events will turn out. Anticipation must "already have disclosed the ambit [*Umkreis*] out of which something can be expected" (GP, 410). The point may, I think, be put this way: that although we may desist from all definite expectations, we do not thereby cancel the future, which would be the case if the future were essentially tied to events: we expect events, when we do, *from* the future, a horizon which does not budge (pass).

Turning now to the original past, the key idea is again that of a time-dimension (ecstasy) prior to the successive common time of things and events. "The past [*Gewesenheit*] of Dasein ought not to be determined on the basis of the common concept of the past" (GP, 411). The adjective *vergangen* and the noun *Vergangenheit* apply in Heidegger's vocabulary only to the common past of things and events, and are best rendered as 'extinct/extinctness' or 'bygone/bygoneness'. "The extinct is that of which we say that it is no more" (GP, 411). "At most, things are extinct" (GP, 412). The contrasting concept of *Gewesenheit*, which applies only to Dasein, may be put into makeshift English as 'alreadiness'. Coined as it is from the past participle *gewesen* of the German verb *sein* = *to be*, *Gewesenheit* aptly hints that Dasein's past is 'still in force', a mode of its Being, a structure of its existence (*Existenz*): "alreadiness is, however, a mode of Being" (GP, 411). Alreadiness is presupposed by extinctness, just as the original future is presupposed by the expectation of events: in retaining something Dasein concomitantly "co-retain[s] its own past" (GP, 275), as the condition of the possibility of some*thing* being retained.

Another way of putting the distinction between the two sorts of past and of future would be to say that Dasein's being is temporal, as distinct from merely in time in the manner of things and events. "A thing that is not temporal, whose being is not determined by temporality, but only occurs within time, cannot be-as-already-being [be *gewesen*], because it does not ex-ist" (GP, 411). Things are in time, a stone is in time, which is to say that a stone does not ex-ist, is not ecstatically self-relating.

Dasein's always-having-already-been is not its past experiences but 'that for the sake of which' it "goes to meet itself in its future" (*kommt auf sich zu*), pro-jects its capacity to be a future totality in which the original privative absence implicit in its alreadiness will be appropriated. In the future Dasein will be 'back' to what is 'was' (which is not its 'first experience': GP, 24, 376, 378, 411). In that sense our past is ahead of us rather than behind us. From which convolutions it is again evident that the dimensions of this original time are not successive.

So much, then, by way of a summary characterization of the fundamental ecstatic temporality of Dasein to which the common world-time is to be reduced. The burden of Heidegger's actual reduction falls on an array of ingenious derivational correspondences between the earlier-considered properties of world-time and the just-considered ecstasies of fundamental time. In our "saying now" (as in 'now the moving body is there') the horizonal *openness* of fundamental time "gets interpreted as" the *publicness* of the world-now (*Offenheit* as *Öffentlichkeit*: GP, 382). Referential significance is grounded in the fact that fundamental temporality is the temporalization of the time as which (not in which) Dasein is concerned for its Being-in-the-world, "for the sake of which it primarily exerts itself" (GP, 383). Every now is a '*time to* . . .' because fundamental temporality is the essence of *ontological concern*.

As for datability: an event accidentally dates a now, but that relation of datedness does not capture what is fundamentally-temporal in the relation of now to events. In uttering (inwardly or outwardly) 'now that/when . . .' the essential dat*ability* of the ecstatic opening of original making-present "speaks itself out" (interprets itself) with concrete reference to a dating event; which is to say that "the now is itself structurally making-present" (GP, 381), thus essentially non-transient like its ground.

The dimensionality of now, finally, "is nothing other than the expression of original temporality itself in its ecstatic character" (GP, 382). Or more fully: "transitional character pertains to every now because temporality as an ecstatic unity is in itself stretched" (GP, 387). Fundamental time is already stretched inasmuch as a given dimension opens a horizon in a dimension outside that given dimension. This dimensionality "carries over" (*miteingeht*, GP, 382) on to the founded level of the world-now, where it shows up as ("lays itself out as") the fact that this now is in its essence transitional between no-more-now and not-yet-now. The structures of fundamental time, namely making-present, making-present-retentively and making-present-anticipatively are re-interpreted as respectively, 'now that . . .', 'at a former now when . . .' and 'at a future now when . . .'.

However, it is a well-known Heideggerian doctrine that Dasein "primarily and usually" exists as "fallen" (*verfallen*) into the midst of the equipmental world it concernfully makes-present: Dasein is "making-present-falling" (SZ, 422) in that it tends to "determine its own being on the basis of the mode of being of occurrent entities" (GP, 384), refract its temporality through the innerworldly time of events. Thus world-time is re-interpreted as ontologically on a par with the transient entities it makes present, as something occurrent and comprised of many transient nows: it is "shorn" (SZ, 427) of the structural properties specific to it as a secondary making-present (after fundamental time), and comes to be thought of as something adhering to events, "especially when it manifests itself in a certain connection with occurrent nature" (GP, 385):

> The nows are therefore also in a way co-occurrent. This means: we encounter entities *and also* nows. Although it is not expressly said that nows are occurrent in the manner of things, they are nevertheless ontologically 'seen' within the horizon of the idea of occurrentness. The nows *pass away*.
>
> (SZ, 423)

> The nows . . . arise and vanish as occurrent entities do, they pass away, just as occurrent entities do, into being no longer occurrent.
>
> (GP, 385)

The result is that world-time (the full now) and its origin in the temporality of Dasein are "covered over" (GP, 422), obscured. And this "common interpretation of world-time as now-time" (SZ, 423) is "reinforced by the way in which it conceptually elaborates its characterization of time" (SZ, 423) – the tensed theory of time is one such elaboration. So considered, as successive, each now is "grasped on the basis of what is still earlier" (SZ, 423); whence, perhaps, the causal theory of time as a further conceptual elaboration.

In Aristotle Heidegger finds evidence both of an appreciation of the non-transient now of world-time, and of the whittling-down of this to now as a "naked" unit of "an occurrent succession" (GP, 385); to something itself eventlike. Aristotle grasps world-time in so far as he "attributes transitional character to the now" and "defines the time in which entities show up as a number encompassing entities" (GP, 387). In a countervailing and prevailing tendency, however, he treats time as a tensed succession of entities: "Aristotle characterizes time primarily as a succession of nows" (GP, 386):

> Aristotle says: Time is . . . something *on* [*an*] motion. But that means: time *is* in some sense. Given that the common understanding of time

> acknowledges Being only in the sense of occurrent being, then time, in so far as it is publicly accessible *with* the movement, 'there', is necessarily something occurrent, be it in objects, in the subject or everywhere. The nows present themselves as within-time. They arrive and vanish, as occurrent entities do.
>
> (GP, 385)

This prevailing entifying tendency is further evidenced when, considering whether time is real or unreal, Aristotle makes the question whether the past and future 'are' or 'are not' turn on whether they are *occurrent entities* – a line of questioning which draws Heidegger's emblematic criticism that Aristotle's thinking here is trapped in the inadequate understanding of Being according to which "to be equals to be occurrent" (GP, 386).

Let us be clear about how the reduction has answered the earlier mentioned threat that the counted nows of motion may be spatialized by the correlated 'heres' which date them. The key thought is that the 'here' is ontologically grounded in the now which makes it present, thus ultimately in the *fundamentally* temporal making-present that is the ecstatic temporality of Dasein. The time that has a quantitative spatial correlate is grounded in a temporality that has no such spatial correlate. Space derives from time:

> Because the *temporality* of facticial being-in-the-world originally *makes possible the opening up of space* and because spatial Dasein has in every case assigned itself to a 'here' on the basis of a 'there', the time that is cared for in the temporality of Dasein is always, in respect of its datability, bound to a place of Dasein. It is not at all the case that time is stuck on to [*geknüpft*] a place, but rather temporality is the condition for the possibility of dating binding itself to the 'spatial-local', and indeed in such a way that the latter is binding for everybody. Time is not coupled with space, but rather the '*space*' which is supposedly coupled *shows up only on the ground of temporality* which cares for time . . . The time made-public in time-measurement is in no way turned into space through dating on the basis of quantity-relationships.
>
> (SZ, 417–418; emphasis added throughout)

Whenever we observe a spatio-temporal process, we make it present in any number of ways, of which measuring it is just one. What is special about the measuring way of making-present (of caring for time) is that it is a making-present "for everyone": "time-measurement accomplishes a pronounced making-present of time, such that by its means what we commonly call 'the time' enters our ken" (SZ, 419). As we saw, the fact that fundamental temporality is originally 'outside of itself' is 'spoken out' at the level of world-time as the publicness of now. What is special about the measuring way of making-present is just that here the speaking out of now as public is strongly *pronounced*, inasmuch as it appeals to "the invariance involved in the idea of a standard . . . available in its constancy for all persons at all times" (SZ, 417). *The* time looms into view when I concernfully make-present some process with specific regard to how many times it contains a publicly inspectable invariant process (e.g. the movement of a secondhand), "without the time-measurement being thematically directed at time itself" (SZ, 417).

As we saw, it is only inasmuch as secondary making-present 'speaks out' (interprets) primary making-present that the 'there' of secondary making-present is temporalized. But this seems to presuppose that primary making-present is itself either not bound to space at all or else bound to a space which is originally temporalized by ecstatic time. For the reduction is thwarted if the temporality to which space-bound common time is to be reduced is itself irreducibly space-bound. But a moment ago we noted Heidegger deny that it was: the spatiality of Dasein, the fact of it always-already having assigned itself to a 'here', rests "on the ground of temporality". What is at stake here for Heidegger is his overarching purpose in SZ (announced in para. 70), namely, the interpretation of all modes of Being as specific styles of temporalization of Dasein's ecstatic temporality: every mode of Being, every "existential", would be ultimately characterizable as Being made-present in some way by time: borne up by a wave of time determinative of the existential's specific essence. Accordingly, the 'is' of space would be temporally made-present, a presentation whose sense would be temporal. Specifically, the making-present of space consists, according to Heidegger, in a "de-distancing" of things (but from where?). Things are spatial inasmuch as de-distanced in the sense of drawn into the ambit of Dasein's concern. Not de-distanced within space, but de-distanced into space, thus de-distanced before all discrimination of metric distances (*Abstände*) within quantitative homogeneous space: "bringing-close, and also the estimation and measurement of distances within de-distanced inner-worldly being-present-to-hand *are grounded in making-present*" (SZ, 369; emphasis added). Within this de-distanced space, nearness and farness are fixed by concern, independently of metrical distance: the spectacles on my nose are normally further away from me than the things I see through them (SZ, 107; GP, 318); the metrical distance between me and a town diminishes as I travel towards it, the original distance (*Entfernung*) goes with me and varies with my impatience to arrive.

Because Dasein's concerns are originally pragmatic, the fact of things being brought close into space is the fact of them being brought 'close to hand', displaying the mode of presence "readiness-to-hand" (*Zuhandensein*). But now the thesis of the temporal *presentedness* of spatial things is in trouble, since the living body (the hand) and the carnal field of manoeuvre into which it extends itself are clearly not presented ("*encounter*ed", *begegnet*). The living body ceases to 'live' (be mine) the moment I attempt to make it appear in either of the modes of presence acknowledged by Heidegger – as ready-to-hand (usable) or as "present-at-hand" (theoretically contemplatable). But this means that, not being made-present, the space of the living body cannot rest on the ground on temporality, which essentially makes-present (lets things stand out *over and against* a subject). Whence it follows that fundamentally space 'is not' (temporally made-present), escapes ontology: spatiality as pragmatic presentedness is derived from a deeper and timeless carnal spatiality irreducible to temporality. Light on how this irreducibility is veiled by Heidegger's text has been shed by the French phenomenologist, Didier Franck, in his important study *Heidegger et le problème de l'espace*,[37] Franck's basic thesis – that Heidegger's project of basing fundamental ontology on temporality founders on the impossibility of reducing carnal spatiality to a mode of temporal making-present – germinates in the long overdue observation that SZ is riven through by this paradox: that while Heidegger's modes of presence, readiness-to-*hand* and presence-at-*hand*, insistently *allude* to a reference of the world to the living flesh, this reference is never properly acknowledged, while such sparse remarks as are offered on the body portray it as a "body-thing" presented to a disembodied Dasein (HPE, 58). Considering Heidegger's remarks on de-distancing:

> de-distancing is proximately and for the most part circumspect bringing-nearer, the bringing-into-proximity by way of procuring, making available, having within reach of the hand. Determinate modes of purely epistemological discovery have, however, also the character of bringing-close:
>
> (SZ, 105)

Franck comments:

> the description shows clearly the transition from one sense of proximity to another. De-distancing is first [described as] taking-in-hand, placing within the reach of the hand. What is close is thus relative to the hand, coincides with the field of manoeuvre and is not a modality of presence, since the flesh incarnates itself without being and without time. Heidegger's endeavour thereafter consists in screening out rather than in reducing [to temporality] this carnal-manual reference in order to assign to proximity, as governed by concern, the temporal meaning of a presentation. In order to do this he severs proximity from what is within the reach of the hand, in affirming that also the theoretical access to entities can be a bringing-close.
>
> (HPE, 82–83)

In other words, if the discovery of purely theoretical entities making no reference to the hand is a case of bringing-close, then there is clearly no essential connection between proximity and the hand. No reduction to temporality of carnal space is actually carried out, but rather the carnal reference is systematically kept out of sight, though it constantly intrudes at the level of Heidegger's manual vocabulary.

As Franck points out in a clinching passage (HPE, 63–64), Heidegger himself acknowledges the a-temporality of the flesh in a late text remarking on the fact that we share 'our' flesh with animals, but not our ecstatic time. The flesh must be "without time" if, as seems reasonable to suppose, the animals who share it with us are bereft of ecstatic temporality. We can 'sympathize' in a Bergsonian sense with a dog stretching,[38] although we know that the dog is not thinking that now it is time to get up. And to cap it all, some thirty-six years after SZ Heidegger himself in his lecture *Zeit und Sein* explicitly disavowed the project of SZ: "the attempt of *Sein und Zeit* to reduce the spatiality of Dasein to temporality is not tenable" (quoted in HPE, 13).

It follows that no reduction of space to time has ensured that a quantity of now–here correlations will not represent a spatialization of time and the 'upward journey' of the concept of time not involve a progressive falsification. The worry was superfluous, however. For Dasein's spatiality, while irreducible to a mode of temporalization of ecstatic time, is also irreducible to geometrical space: "the spatiality of Dasein is not determined by the indication of a position where a corporeal being is present-to-hand" (SZ, 107). 'Here' is not a point in space indifferently occupied by my body, but "the space of manoeuvre" of Dasein's "de-distancing being-alongside innerworldly entities" (SZ, 107).

Fundamentalness, objectivity, realness

I have reconstructed Heidegger's position as follows. In order to protect the ontological primacy of phenomenological time against the replacement thesis, it is not necessary to deny, as Bergson does, the reality of time-quantities; it suffices to establish that the quantification of time is a genuine, though derived, mode of temporalization of fundamental time. That is, it suffices to show that the scientific-quantitative construction of nature is grounded in Dasein's fundamental time and world. In Heidegger's words:

> The elucidation, based on the temporality of Dasein, of the ontological foundations of [time-] measurement presupposes a clarification of world-time and of within-timeness, as well as an elucidation of the existential-temporal constitution of the discovery of nature and of the temporal meaning of measurement as such. An axiomatics of physical measurement techniques *rests on* these investigations and is incapable on its side of ever unpacking the time-problem.
>
> (SZ, 417–418, footnote; emphasis added)

We may construe this as endorsing the proposition we earlier noted Merleau-Ponty assert, namely, that because fundamental time is pre-understood to scientific conceptions of time, the latter cannot *dispense with* the former: it is true that science would not get off the ground if we did not have an understanding of time prior to operational definitions of it. But from this by itself the conclusion the phenomenologists want – that phenomenological time is real time, *irreplaceable* – cannot be validly derived. The history of philosophy teems with denunciations of 'the confused perceptions of the senses', conjoined with the injunction that our immediate construals of concepts should *give way to* a derived constructive-philosophical or scientific understanding. A replacement theorist need not deny the indispensability of the first understanding to which the phenomenologists appeal. What he or she would deny is that 'the ladder cannot be kicked away'. To secure their case against the replacement thesis the phenomenologists need to prove that the time identified as fundamental is also irrevisably *real* time. Short of this there would seem to be no reason why the honorific adjectives 'fundamental' and 'original' should not be taken as referring simply to a temporality that is first in the hierarchical order of 'merely subjective' temporalities.

We are not helped by the fact that the obscurely honorific notion of real time has not been a matter of debate among phenomenologists post-Bergson, while in analytical quarters it has been much bandied about without being subjected to the scrutiny it deserves. What is real about real time? Is this asking after general 'criteria for realness'? I shall argue that such criteria as might be mustered from the literature evaporate under our gaze, and that the only chance of answering the title question of this chapter lies with ditching the criterial approach and rethinking the question as a request for elucidation of the "sense of Being" (*Seinssinn*) suggestive, in Heidegger's early philosophy, of a nexus between time and Being.

A positive answer might seem to be suggested by the phenomenologists' denial that time inheres in physical reality. Given this denial, might it not be claimed that phenomenological time is real time on the ground that there exists no rival time to complete for the title? By elimination, then, phenomenological time would be 'as real as real can be'. This answer is wholly unacceptable, since the question whether there are rival times is settled positively by the fact that rival claims are made, not negatively by the phenomenological rejection of natural time.

Many philosophers have argued that tenseless time is real time, and tensed time unreal or (some say) subjective; and although they do not come clean on their notion of realness, their arguments tacitly espouse criteria for the same. One of these is 'basicness to understanding', as when the justification offered for the realness of tenseless time is that tenseless token-reflexive truth-conditions give the meaning of tensed sentences, whereas tensed truth-conditions do not give the meaning of tenseless sentences: everything we might say in tensed style could be said in tenseless style, but not vice versa. But it is not clear why this basicness to thought has a better claim to hallmark realness rather than, say, 'being true of a mind-independent physical reality'. Why is the understanding said to be basic not merely subjective, for all its basicness? What connects it with mind-independent reality?

Many tenseless theorists operate with both criteria, mixing without matching.

Others have buttressed the thesis that tenseless time is real *qua* conceptually basic with the thesis that it is real *qua* rational. The two theses tend to occur together in the context of pointing out that tenseless time is not infected by the self-contradictoriness of tensed time, as alleged by McTaggart. The self-contradictoriness of tensed time, however, is not regarded as depriving it of *all* reality, but rather as relegating it to a degenerate existence "in our heads".[39] It would seem that tensed time can 'be' despite its self-contradictoriness, as long as it remains inside our heads; outside them it is flatly unreal. The criterion operating here under the guise of the requirement that real time be rational is at bottom the requirement that it be mind-independently objective; which leaves phenomenological time no chance of being real time.

The criterion just considered applied the principle that something is real *because* it is mind-independently objective. From a Heideggerian perspective that is false: something (a mode of presence) that is real can be real *as* (Dasein-dependently) objective. Realness is not co-extensive with the objectivity of occurrent entities. Realness is what an entity possesses inasmuch as it derives from Being, not inasmuch as it is an entity of a kind different from Dasein. What is real as pertaining to Dasein is no less real than what is real as objective.

'Real as objective' – that 'as' alludes to a Heideggerian sense. Everything is in one way or another 'real as . . .', even fictions. Husserl claimed that all senses are constituted by and grounded in consciousness, and that they do not entail the mind-independent existence of the objects exhibiting them. For Heidegger, by contrast, senses are ultimately grounded not in consciousness but in Being: every sense is in one way or another a "sense of Being". Objectivity and subjectivity are senses of Being arising within Dasein's world, which, being prior to that distinction, cannot properly be called objective or subjective. But the point crucial to my problematics here is the question how time connects with Being via senses of Being.

I think it will help ride to earth Heidegger's notion of sense of Being if at this point we offset his 'concept' of the Being/beings relation against Sartre's. Imagine a slab of paste with circles impressed in it. Let the paste stand for the Being of beings and the circular areas for phenomenal beings cut out within it by perceptual and/or speaking intentionality. We now have an image, sufficient for present purposes, of Sartre's version of the Being/beings relation as that holding between an undifferentiated "transphenomenal being-in-itself"[40] and the discrete phenomena sharing in it. Notice that this relation is such that Being stands to beings as a whole to its parts and that the whole is made of the *same* stuff ('paste') as its parts. Because there is nothing like Heidegger's ontological difference here, it is impossible that a being should bear a sense of its Being. The only transaction between Sartre's Being and his beings is a tendency, corresponding to a flagging of intentionality, for the latter to collapse back into the transphenomenal paste out of which they are phenomenalized. Being (Being) is not here something of which beings can present a sense, for this sense inhabits ontological *difference*. The point here is not that Sartre's beings are senseless in the sense of existing without reason (as he insists) but that they do not possess the sense of a gift (as Heidegger puts it, playing with *es gibt*) from a Be-ing which effaces itself in the 'act' of offering it, leaving phenomena vibrating with the sense of that Be-ing which prompts the question "why [there is] something rather than nothing".[41] I understand it to have been Heidegger's idea that this sense is fundamentally temporal, that it inhabits time as making-present, that making-present is the making of the ontological difference.

As is well known, SZ remained unfinished because Heidegger postponed indefinitely, if he did not actually abandon, his project of thinking through the connection between time and Being. Whatever impasses may have blocked his way forward, the suggestions in SZ of some sort of pact between time and Being (the status of time as the "vestibule of Being" or the "horizon for Being": SZ, 437) remain, I believe, of irrefragable value as the only possible approach towards justifying the claim that phenomenological time is real time.

Notes

1 Time as creative process: Bergson

1 References in the text to Bergson's works are to the Presses Universitaires de France editions (Paris). Titles are abbreviated as follows:

DI	*Essai sur les données immédiates de la conscience* (1889) 1967
MM	*Matière et Mémoire* (1896) 1968
EC	*L'Evolution Créatrice* (1907) 1969
DS	*Durée et Simultanéité* (1922) 1968
DSMR	*Les Deux Sources de la Morale et de la Religion* (1932) 1973
PM	*La Pensée et le Mouvant* (1934) 1969

2 F. Copleston, *A History of Philosophy*, vol. ix, *Maine de Biran to Sartre*, London: Search Press, 1975, p. 179.

3 Imprecision in the sense of a tendency not to complete analyses is clearly brought out by A.R. Lacey in his careful and patient *Bergson*, London: Routledge, 1993.

4 Friends of Bergson anxious to exempt him from the 'critique of presence' have pointed out that some of his remarks on the present suggest he thinks the present is never really present. Thus, Constantin Boundas:

> It is as if the present is never really present to itself, haunted as it is by the past that it is in the process of becoming and by the future that it is in the mode of not yet being it. Present and past alike are haunted by an immemorial past which has never been.
>
> ('Deleuze–Bergson: an ontology of the virtual', 1996, p. 101)

I don't think this lets Bergson off the hook, since it does not touch the crucial issue (to be pursued in this chapter) whether he has a principle of temporal difference in virtue of which the present *can* be a passing into non-presence (rather than a stable presence). The issue of the principle of temporal difference is particularly acute here owing to Bergson's rejection of all negative forms of temporal structuring. Given that rejection, it cannot really be said that Bergson's present 'is its future in the mode of *not* yet being it': that expression comes from Sartre, for whom temporal difference is emphatically negative – "nihilation". For exemptions from the critique of presence on behalf of Sartre and Merleau-Ponty, see respectively Christina Howells, 'Qui perd gagne: Sartre and Derrida', *The Journal of the British Society for Phenomenology*, 1982, vol. 13, pp. 26–34; and Mark Yount, 'Two reversibilities: Merleau-Ponty and Derrida', *Philosophy Today*, 1990, vol. 34, pp. 129–140.

5 G. Deleuze, *Le Bergsonisme*, Paris: Presses Universitaires de France, 1966, p. 41.

6 Levinas in interview. R. Kearney, *Dialogues with Contemporary Continental Thinkers*, Manchester: Manchester University Press, 1984, pp. 49–50.
7 M. Merleau-Ponty, *Eloge de la Philosophie*, Paris: Gallimard, 1960, pp. 297–298.
8 Ibid., p. 293.
9 The issue of how far the phenomenologists acknowledged their indisputable debt is decided to the latters' shame by, among others, Leszek Kolakowski: 'French Existentialist philosophy was also Bergson's heir, usually without acknowledging the debt' (*Bergson*, Indiana: St. Augustine's Press, 1985, p. 102).
10 "Bergson, our contemporary, is back" announces the blurb of a recent collection of essays: *The New Bergson*, J. Mullarkey (ed.), Manchester: Manchester University Press, 1999.
11 F.C.T. Moore, for one, has invoked this one-way independence to endorse Bergson's account of internal time-experience while judging as "dubious" his "attempt to create a sort of super-phenomenology for life itself, analogous to a phenomenology for an individual consciousness". *Bergson: Thinking Backwards*, Cambridge: Cambridge University Press, 1996, p. 9.
12 This point has been made by Lacey, apropos of a cricket ball: *Bergson*, p. 97.
13 J.-P. Sartre, *L'Etre et le Néant*, Paris: Gallimard, 1943, p. 235.
14 See especially Bertrand Russell, *The Analysis of Mind*, London: Allen & Unwin, 1971, pp. 145–148.
15 See M. Shorter, "Imagination", *Mind*, 1958, vol. lxi, p. 542.
16 Deleuze, *Le Bergsonisme*, p. 17.
17 Given that all this is subject to the overarching cosmological dimension of Bergson's philosophy, centred on the notion of the self-differentiating "surge of life", the question arises whether *ultimately* there are many *durées* or just one. Into this debate I shall not enter since it takes us too far from phenomenological time. For an interpretation favouring just one, see Deleuze, *Le Bergsonisme*, Chapter IV.
18 Suggestive though it is, Bergson's notion of temporal rhythm is vague and beset with problems he seems not even to suspect. On, for example, the difficulty of finding a yardstick for comparing rhythms, see Lacey, *Bergson* p. 129.
19 G. Deleuze, 'La conception de la différence chez Bergson', *Les Etudes bergsoniennes*, 1956, vol. 4, p. 88. See also p. 103, "what differs is difference itself"; p. 88, "what differs has itself become a substance". The whole of this commentary rests on Deleuze's seminal idea that difference precedes and conditions sameness/repetition.
20 B. Russell, *History of Western Philosophy*, London: Allen & Unwin, 1961, p. 763.
21 Russell, *The Analysis of Mind*, pp. 159–160.
22 Russell, *History of Western Philosophy*, p. 764.
23 R. Gale, 'Some metaphysical statements about time', *The Philosophical Review*, 1966, vol. lii, p. 326.
24 R. Gale in R. Gale (ed.), *The Philosophy of Time*, London and New York: Macmillan, 1968, p. 389.
25 Sartre, *L'Etre et le Néant*, p. 156. To avoid excessive notes, this work will be referred to as 'EN' throughout the remainder of this chapter.
26 C.D. Broad makes the same point when he says of John McTaggart in his charming monograph that he "unwittingly exemplified Bergsonian principles by performing actions and expressing opinions which were incalculable before the event but rationally explicable after it", *Ethics and the History of Philosophy*, London: Routledge & Kegan Paul, 1952, p. 90. To my knowledge, Sartre and Broad are the only writers to have seen the centrality of this principle to Bergson's philosophy.
27 A number of commentators have claimed that *durée* has no structure at all, or none that is sayable (see, for example, Kolakowski, "the inexpressible experience

of time", *Bergson*, p. 87). While this view is justifiable, I decline it in favour of the structured interpretation offered above, both because I believe it textually is better justified and because it enables us to bring Bergson into critical contact with the phenomenologists: to hold the structureless view is to enjoin silence.

28 Deleuze, *Logique du Sens*, Paris: Les Editions de Minuit, 1968, p. 302.

29 J. Mullarkey, *Bergson and Philosophy*, Edinburgh: Edinburgh University Press, 1999, p. 141.

30 Deleuze, *Logique du Sens*, p. 203.

31 Merleau-Ponty, *La Phénoménologie de la Perception*, Paris: Gallimard, 1945, p. 319.

32 Ibid., p. 429.

33 Merleau-Ponty, *L'Union de l'âme et du corps chez Malbranche, Biran et Bergson*, Paris: Vrin, 1968, p. 95.

34 See Husserl: "the sort of combination uniting consciousness with itself can be characterized as *synthesis*, a mode of combination exclusively peculiar to consciousness", *Cartesianische Meditationen und Parier Vorträge*, The Hague: Martinus Nijhoff, 1973, p. 39.

35 Merleau-Ponty, *La Phénoménologie de la Perception*, p. 319.

36 Deleuze, *Le Bergsonisme*, p. 23.

37 Ibid., p. 57.

38 Ibid., p. 23.

39 Merleau-Ponty, *La Phénoménologie de la Perception*, p. 320.

40 Ibid.

41 As Heidegger famously put it.

42 At this point friends of Bergson are prone to reply that we don't look hard enough for 'evolution', or that the pragmatic orientation of our perception constrains us to carve out artificial fixities from the flux. Thus Mullarkey commenting on Lacey's invariant cricket ball: "just because our mundane powers of perception are normally closed to what is ongoing in our environment, it does not follow that there is nothing there" (*Bergson and Philosophy,* pp. 140–141). This is argument *ex ignoratia*, a plea for an act of faith, using words typically uttered by believers in ghosts trying to convert non-believers.

43 J.-P. Sartre, *La Transcendance de l'Ego*, Paris: Vrin, 1972, p. 46.

44 See the discussion of 'all-at-onceness-with-respect-to-extension' in relation to the dimensions of Husserl's absolute consciousness, pp. 46–47 *et seq*.

45 Sartre, *La Transcendance de l'Ego*, p. 63.

46 "Neither Bergson nor the psychologists he criticizes distinguish consciousness from the object of consciousness". Merleau-Ponty, *L'Union*, p. 81.

47 It certainly cannot be dismissed by the question-begging assertion that "the objections of Sartre and Merleau-Ponty are too much involved in the phenomenology of consciousness and subjectivity to be of much use to the theory of real time" (Boundas, 'Deleuze–Bergson', p. 101). The extremely complex issue of real time is taken up in Chapter 6.

48 Kolakowski, *Bergson*, pp. 103–105.

2 Time as time-consciousness: Husserl

1 *Zur Phänomenologie des inneren Zeitbewusstseins (1893–1917)*, The Hague: Martinus Nijhoff, 1966. Henceforth 'ZB'.

2 See, in particular, Rudolf Bernet's Introduction to the German 'student' edition of Part B of ZB: *Texte zur Phänomenologie des inneren Zeitbewusstseins (1893–1917)*, Hamburg: Felix Meiner Verlag, 1985, pp. x–lxxiii. And by John Brough: 'The emergence of an absolute consciousness in Husserl's early writings on time-consciousness', in F. Elliston and P. McCormick (eds), *Husserl: Expositions and*

Appraisals, Notre Dame: University of Notre Dame Press, 1977, pp. 83–100. Also Brough's excellent Introduction to his translation of ZB: *Edmund Husserl: On the Phenomenology of the Consciousness of Internal Time*, London: Kluwer Academic Publishers, 1991, pp. xi–lvii.

3 *Ideen zu einer reinen Phänomenologie und phänomenologischen Philosophie* (Erstes Buch), The Hague: Martinus Nijhoff, 1950, p. 197 (abbreviated: *Ideen I*).

4 *Cartesianische Meditationen und Pariser Vorträge*, The Hague: Martinus Nijhoff, 1973, p. 23.

5 The notion that things are time-things has currency in the Anglo-American tenseless theory of time, where *this* substance is analysed as the *unique* series of causally-connected tenseless events comprising its 'history'. The piquancy of Husserl's version of time-thing is that the time-relations into which 'substance vanishes' are relations of neither tenseless nor tensed time.

6 As R. Sokolowski has noted:

> Husserl's theory of constitution has nothing to say about the material content of things: it explains the constitution of temporality but . . . does not explain how one object in time is constituted as a man, another as a stone, a third as a painting.
>
> (*The Formation of Husserl's Concept of Constitution*, 1970, pp. 109–110)

7 This extent is not the 'specious present' for which it has been mistaken by many of Husserl's commentators. The specious present is the time-structure of the sensory field, whereas an extent is a stretch of the unified *consciousness of* (constituting) that field, as has been well shown by Shaun Gallagher in his *The Inordinance of Time*, Evanston, IL: Northwestern University Press, 1998, pp. 17–52.

8 This diagram is courtesy of R.M. Zaner (*The Problem of Embodiment*, The Hague: Martinus Nijhoff, 1971, p. 225), who says that it derives from Husserl via an expansion by his pupil Dorian Cairns. Unfortunately Zaner does not indicate whether it is expanded from any of the diagrams appearing in ZB, though the one on p. 330 looks like a strong candidate. It should be borne in mind that Husserl was not satisfied with any of his diagrams.

9 For a rival account of these two dimensions of co-consciousness, see Barry Dainton *Stream of Consciousness*, London: Routledge, 2000, especially p. 168.

10 *Cartesianische Meditationen*, pp. 80–81.

11 This Bergsonian misreading is made by the unBergsonian Merleau-Ponty on pp. 477–479 of *La Phénoménologie de la Perception*, Paris: Gallimard, 1945. For a commentary on this, and on Merleau-Ponty's misunderstanding of synthesis of identification, see Zaner, *The Problem of Embodiment*, pp. 225–230.

12 *En Découvrant l'Existence avec Husserl et Heidegger*, Paris: Vrin, 1974, p. 156.

13 *La Voix et le Phénomène*, Paris: Presses Universitaires de France, 1967, p. 72.

14 Ibid., p. 95.

15 Bernet aptly compares Husserl's consciousness to a ruminant that ingests endlessly and excretes nothing (*Texte zur Phänomenologie*, p. liv). See also Dainton (*stream of Consciousness*, pp. 156–158) on Husserl's "clogging of consciousness"; also David Wood: "there is little sense here of what might be called the negative side of time, the threat it poses to all constitution of identity" (*The Deconstruction of Time*, Evanston, IL: Northwestern University Press, 2001, p. 103).

16 Brough, Introduction to *Edmund Husserl*, p. lv.

17 Gallagher, *The Inordinance of Time*, p. 194. See also David Wood, *The Deconstruction of Time*, pp. 358–359. Gallagher and Wood argue that Husserl's 'lived' time is already impregnated with the reflective and non-linear structures of narrative. For further criticism of Husserl's serial bias see the excellent study by R. Duval: 'La durée et l'absence', *Revue des Sciences Philosophiques et Théologiques*, 1981, vol. 65, pp. 521–572.

18 As Merleau-Ponty indicated in an unfinished late work, Husserl fails to consider "the influence of 'contents' on time . . ., of *Zeitmaterie* on *Zeitform*", *Le Visible et l'Invisible*, Paris: Gallimard, 1964, p. 238.
19 In *Temps et Récit* (vol. iii, Paris: Editions du Seuil, 1985) Paul Ricoeur argues that "the appearance of the talk of form" (p. 75) in Husserl initiates a series of illicit borrowings from Kantian time which produce paradoxes or "aporias" in the phenomenological setting. I agree, with the reservation that I think Ricoeur tends to exaggerate the degree of paradox. It is, for example, odd – especially for an interpreter of analytical philosophy to a phenomenological readership – to speak of "rediscovering [in Husserl] the paradox of Kant according to which time itself does not pass" (Ricoeur, p. 453). This disregards: (i) the fact that Husserl's non-passing time-relations are *constituted by* passing time, what is not the case with Kant; and (ii) that what almost all analytical positions find paradoxical is the notion that time (as distinct from events) *does* pass: see Chapter 4.
20 *L'Etre et le Néant*, Paris: Gallimard, 1943, p. 156.
21 "Husserl ne pense pas la primitivité du paraître", as G. Granel says untranslatably in his unsurpassed Heideggerian critique of Husserl: *Le Sens du Temps et de la Perception chez E. Husserl*, Paris: Gallimard, 1968, p. 242. Granel claims further, however, (ibid., pp. 117–118) that Husserl's picture of time is a fiction produced by the ontic commitments of his descriptive language. Being too impressed by the reality of "the wonder of time-consciousness" to agree, I would say instead that Husserl's time may be the product of reflection. This difference leaves me in full agreement with Granel's important and well-demonstrated thesis that Husserl is insensitive to "the obstinacy with which Being infiltrates itself" (ibid., p. 265).
22 *L'Etre et le Néant*, p. 713.
23 Brough, Introduction to *On the Phenomenology of the Consciousness of Internal Time*, p. lv (my emphasis).
24 Bernet, 'Introduction', *Texte zur Phänomenologie*, p. lvi; emphasis original.
25 *Ideen I*, p. 115.
26 As by Merleau-Ponty in his late writings: "this perceptual world is at bottom Being in the sense of Heidegger", *Le Visible et l'Invisible*, p. 223.

3 Time beyond being: Levinas

1 The following works by Levinas are abbreviated in this chapter as follows:

EE	*De l'Existence à l'Existant*, Paris: Fontaine, (1947*) 1978
TA	*Le Temps et l'Autre*, Paris: Presses Universitaires de France, (1979) 1989*
TI	*Totalité et Infini*, The Hague: Martinus Nijhoff, (1961) 1980*
AEE	*Autrement qu'Etre ou au-delà de l'Essence*, The Hague: Martinus Nijhoff, 1978
TIPH	*Théorie de l'Intuition dans la Phénoménologie de Husserl*, Paris: Vrin, (1930) 1970*

Page numbers refer to the asterisked editions.
2 *en deçà de*. Note that Levinas makes synonymous metaphorical use of two spatial expressions opposite in literal meaning: *au delà de* (beyond) and *en deçà de* (this side of, hither side of).
3 Levinas, 'La réalité et son ombre', *Les Temps Modernes*, 1948, vol. 38, p. 776.
4 As Stephan Strasser has rightly remarked (*Jenseits von Sein und Zeit*, The Hague: Martinus Nijhoff, 1978, p. 374), Heidegger explicitly rejects this view of Being on p. 3 of *Sein und Zeit*. See also Edith Wyschogrod, *Emmanuel Levinas: The Problem of Ethical Metaphysics*, The Hague: Martinus Nijhoff, 1974, p. 11.

5 Levinas later reiterated the charge of totalization apropos of Heidegger's interpretation of Dasein as history: "when Heidegger interprets our being-in-the-world as history . . . Dasein is its history to the extent that it can interpret and narrate its existence as a *finite* and contemporaneous story, a totalizing co-presence of past present and future", *Dialogues with Contemporary Continental Thinkers*, Richard Kearney (ed.), Manchester: Manchester University Press, 1984, p. 56.
6 *Was ist Metaphysik?*, Frankfurt: Klostermann, 1986, p. 31.
7 Note that Levinas uses the following terms synonymously: *l'il y a* (the there is), *l'exister* (the existing), *l'existence* (existence), *le verbe être* (Be-ing), all of which contrast with *l'existant* (the human subject) and with *l'existent* (a determinate entity or being). The symmetry with Heidegger is patent: the *il y a* corresponds to Heidegger's *Sein*, while *existent* and the *existant* correspond respectively to *Seiendes* and to *Dasein*. These correspondences may explain Levinas's remark that Heidegger's ontological difference is "for me the deepest thing in *Sein und Zeit*" (TA, 24).
8 Romanticism turned this on its head: in Lamartine's injunction "aimez ce que jamais vous ne verrez deux fois" (*Le Lac*) it is the instant, not everlastingness, that simulates eternity.
9 Levinas, "Vérité de dévoilement et vérité de témoignage", *Archivio di Filosofia*, Rome: 1972, pp. 105–106.
10 *En Découvrant l'Existence avec Husserl et Heidegger*, Paris: Vrin, 1974, p. 155.
11 Something of the flavour of this troubled light may be got from the "black sun" motif of Albert Camus' early essays, 'L'envers et l'endroit' and 'Noces á Tipasa', in *Essais*, Paris: Pléiade, pp. 15–60.
12 As predicated of enjoyment, 'intentional', 'constitution' etc. are makeshift placeholders for a yet-to-be-invented language, since they poach on the vocabulary of the Husserlian theoretic intentionality with which enjoyment is being contrasted. Of this Levinas issues sufficient warning: "if we could still speak of constitution here, we should . . ." (TI, 101); "its intention, if it is permissible to use that term . . ." (TI, 109, see also, 95, 98). The intentionality of enjoyment retains the idea, germane to Husserlian intentionality, of the transitive directedness of consciousness. Husserl, however, does not consider this directedness a sufficient determination of intentionality. The essence of intentionality lies for him in the noetic-noematic correlation characterizing transitive directedness.
13 Thus Sartre, for one, pokes fun at the "digestive philosophies of empirico-criticism and neo-Kantianism", and the "alimentary philosophy" of idealism in 'Une idée fondamentale de la phénoménologie de Husserl, l'intentionnalité', *La Nouvelle Revue Française*, 1939, p. 129.
14 Levinas, *De Dieu qui Vient à l'Idée*, Paris: Vrin, 1992, p. 161 (emphasis original).
15 Levinas, 'Vérité de dévoilement et vérité de témoignage', p. 104.
16 This is a typical case of one phenomenologist distancing himself from another rather than criticizing him. As Derrida noted, "Levinas respects the zone or level of traditional truth, and the philosophies whose presuppositions he describes are in general neither refuted nor criticized" (*L'Ecriture et la Différance*, Paris: Editions du Seuil, 1967, p. 137). This is surely truer of Levinas on Husserl than on Heidegger, who does not work at the level of traditional truth.
17 *Sein und Zeit*, Tübingen: Max Niemeyer, 1986, p. 212.
18 Ibid., pp. 211–212.
19 Ibid., p. 212 (emphasis added).
20 Ibid.

4 The language of time

1 *L'Evolution créatrice*, Paris: Presses Universitaires de France, 1969, p. 2.
2 *Zur Phänomenologie des inneren Zeitbewusstseins (1893–1917),* The Hague:

Martinus Nijhoff, 1966, p. 74. Further references in the text to this work are abbreviated as 'ZB'.

3 *L'Etre et le Néant*, Paris: Gallimard, 1943, p. 190.

4 *La Phénoménologie de la Perception,* Paris: Gallimard, 1945, p. 484.

5 *De l'Existence à l'Existant*, Paris: Fontaine, 1947, p. 125.

6 See in particular J.J.C. Smart, 'The river of time', *Mind*, 1949, vol. 58, 483–494.

7 A.S. Eddington, *Space, Time, and Gravitation*, Cambridge: Cambridge University Press, 1953, p. 51.

8 *The Language of Time*, London and New York: Humanities Press, 1968, p. 230. Further references to this book will be abbreviated to 'LT' plus page number. (Gale has recently converted to tenseless theory.)

9 In other words, the change of time is presupposed by change in time. See C.D. Broad, *Scientific Thought*, London: Routledge & Kegan Paul, 1969 (1923), pp. 67–68.

10 I have been alluding here to A.N. Prior's logicizing proposals for analysing tense without reference to becoming. Further discussion and criticism of this mean-and-lean approach to tense are offered in Chapter 5, pp. 151–152.

11 K. Sneddon, *Time*, London: Croom Helm, 1987, p. 5.

12 Ibid.

13 *Ethics and the History of Philosophy*, London: Routledge & Kegan Paul, 1952, p. 213.

14 *La Voix et le Phénomène*, Paris: Presses Universitaires de France , 1967, p. 70.

15 'La réalité et son ombre', *Les Temps Modernes*, 1948, no. 38, p. 779.

16 Merleau-Ponty, *La Phénoménologie de la Perception*, p. 277. The theme of the absolute past originates with Merleau-Ponty and not, as commonly supposed, with Levinas.

17 D.C. Williams, 'The myth of passage', in R. Gale (ed.), *The Philosophy of Time*, London: Humanities Press, 1968, p. 105.

18 *Papers on Time and Tense*, Oxford: Oxford University Press, 1968, p. 28.

19 P.J. Zwart, 'The flow of time', *Synthese*, 1972, vol. 24, p. 135.

20 E. Harris, 'Time and change', *Mind*, 1957, vol. lxvi, p. 234.

21 *L'Etre et le Néant*, Paris: Gallimard, 1943, p. 702. Henceforth abbreviated in this chapter to 'EN'.

22 *La Phénoménologie de la Perception*, p. 482.

23 Ibid.

24 Russell countered Bergson's famous claim that real motion is comprised of motions (not of immobilities) with the assertion that it commits the fallacy of composition, by assuming that an unanalysable whole must be made up of unanalysable wholes. A friendship is an unanalysable whole that is not, in Russell's view, composed of friendships, but of distinct friends (*History of Western Philosophy,* p. 833). So by parity of reasoning, the indivisibility of motion does not entail that motion cannot be composed of a body visiting discrete places at discrete moments of mathematical time. Russell's objection is toothless because he does nothing to show that Bergson's view of motion is *a case* of the fallacy. (It is not even obvious that it would be committed by someone who supposed that a friendship is composed of friendships. Might a friendship not be composed of irreducibly social acts of friendship? Perhaps it all depends on context: a football team lined up for a photograph is composed of players, but is that still true of the team in full fluid action?)

25 *La Phénoménologie de la Perception*, p. x.

26 Augustine, *Confessions*, Oxford: Oxford University Press, 1992, p. 230. For Wittgenstein's treatment of the Augustinian predicament, see his *Philosophical Investigations*, Oxford: Basil Blackwell, 1981, p. 79; also *The Blue and Brown*

Books, Oxford: Basil Blackwell, 1975, pp. 108–109. For endorsements and developments of this line of criticism see R. Suter, 'Augustine on Time with some Criticisms from Wittgenstein', *Revue Internationale de Philosophie*, 1961–1962, pp. 378–389.

27 *Ideen zu einer reinen Phänomenologie und phänomenologischen Philosophie, Erstes Buch,* The Hague: Martinus Nijhoff, 1950, p. 86 (abbreviated as *Ideen I*).

28 *Ideen I*, p. 85; emphasis added.

29 *Eloge de la Philosophie*, Paris: Gallimard, 1960, p. 242.

30 *La Phénoménologie de la Perception*, p. ix.

5 McTaggart and Husserl

1 The argument first appeared in *Mind*, 1908, vol. 68, pp. 457–474. A later and in some ways fuller version is included in McTaggart's *The Nature of Existence*, Cambridge: Cambridge University Press, 1927, vol. 2, pp. 9–31. References in the text will be to both versions, abbreviated respectively as 'M' and 'NE'. To readers familiar with the argument my apologies for reproducing it here for the *n*th time. Of the two evils of doing so and not doing so, the former seemed the lesser.

2 For a complete list of the candidates for change rejected by McTaggart, see M, 459–460.

3 K. Sneddon, *Time,* London: Croom Helm, 1987, pp. 47–48.

4 A.J. Flew, 'The sources of serialism', in S.C. Thakur (ed.), *Philosophy and Psychical Research*, London: Allen & Unwin, 1976, pp. 86–87.

5 D.F. Pears, 'Time, truth and inference', *Proceedings of the Aristotelian Society*, 1950–1951, vol. lii, p. 7. See also F. Christensen, 'McTaggart's paradox and the nature of time', *The Philosophical Quarterly*, 1974, vol. 24, pp. 289–299.

6 'The largest block capitals in the world would not over-emphasize this mistake of McTaggart's', Pears, 'Time, truth', p. 6.

7 For a fairly comprehensive review of McTaggart's use of tenses, see Kenneth Rankin, 'McTaggart's paradox: two parodies", *Philosophy*, 1981, vol. 56, pp. 333–348, especially pp. 343–344.

8 Ibid., p. 346.

9 This is how it is taken by, for example, D.H. Mellor in his *Real Time*, Cambridge: Cambridge University Press, 1984, p. 98.

10 Rankin, 'McTaggart's paradox', p. 346.

11 A.N. Prior, *Papers on Time and Tense*, Oxford: Oxford University Press, 1968, p. 21.

12 Ibid., p. 3.

13 Ibid., p. 20.

14 Ibid., p. 21.

15 Ibid.

16 Prior, 'The notion of the present', *Studium Generale*, 1970, vol. 23, p. 245.

17 This point has been well put by Geneviève Lloyd:

> Mr N.N. cannot, for Prior, be said to be dead. Death does not merely deprive Mr N.N. of present actuality while leaving him, *qua* subject, intact. It deprives us not just of his company. Being dead he is altogether absent from the order of things, just as he was before he came to be.
>
> ('Time and existence', *Philosophy*, 1978, vol. 53, p. 227)

Note that this is a deeper point than the one made by Wittgenstein in *Philosophical Investigations*, I, 40. Wittgenstein's point is that if names got their meanings from their bearers, the name 'Mr N.N.' would cease to have a referent upon

Mr N.N.'s demise, whence it would follow that it could not be used to *say* that he is dead. Whereas Lloyd's point is that on Prior's showing he cannot *be* dead (and for that reason cannot be said to be dead either).

18 As D.H. Mellor has pointed out, *Real Time*, p. 98.

19 Rankin, 'McTaggart's paradox', p. 338; my emphasis.

20 This position is taken by Mellor (*Real Time*, pp. 92–102), by Sneddon (*Time*, pp. 44–60) and, famously, by Russell.

21 See Mellor, *Real Time*, pp. 75ff. Note that the specification of a B-date is essential to this reductivist strategy. A sentence like "Rain is earlier by one week than this token of 'It rained'" is still tainted with tense, since *this* means 'the one I am uttering or inscribing *now*'. This need for a tenseless B-date makes the token-reflexive theory of tense look thoroughly implausible. It is of course true that anyone who understands 'It is raining' understands that utterance to be true and only if it is simultaneous with rain. But the artificiality of appending dates seems to me to argue that the simultaneity is rightly understood as 'sharing the same present' not as 'sharing the same tenseless date'.

22 Quoted by McTaggart (from *Principles of Mathematics*, sec. 422), NE, p. 14.

23 "of which they are states": this addendum shows that McTaggart consciously rejects the analysis (associated with Quine *et al.*) of substance into a series of events as 'thing-stages', tied together by a causal law unique to the series and comprising the becomingless 'history' of the substance.

24 C.D. Broad, *Scientific Thought*, London: Humanities Press, 1969 (1923), pp. 57–58.

25 Notice the ambiguity of McTaggart's introductory remark: "Positions in time . . . are distinguished in two ways" (M, 458; NE, 9). Two ways of distinguishing positions in *one* series, or *two* series counter-distinguished by different types of relation between terms? The ambiguity is perpetuated by commentators who talk of two 'descriptions' etc. of time-positions. The trouble with persisting in this 'linguistic' mode is that it suppresses the paramount ontological issue whether the two descriptions refer respectively to two types of time-series that are really different, or just separated by abstraction. In either case there are, of course, two descriptions.

26 The death of Queen Anne might be intersubjectively constituted as an event in historical time, but to show this would require extending the constitutional project far beyond the sphere of Husserl's micro-durations. This is not to suggest that Husserl was uninterested in historical time. On the contrary, his concern with it deepened steadily from the early 1920s onwards. But he never achieved anything approaching a systematic understanding of the *connection* between phenomenological and historical time, as has been shown by Elizabeth Ströker in her "Zeit und Geschichte in Husserl's Phänomenologie: zur Frage ihres Zusammenhanges", in E.W. Orth (ed.), *Zeit und Zeitlichkeit bei Husserl und Heidegger*, Freiberg: Verlag Karl Alber, 1983, pp. 111–137.

27 I borrow this example from Sartre: *L'Imagination*, Paris: Presses Universitaires de France, 1969 (1936), p. 98.

28 *Ideen zu einer reinen Phänomenologie und phänomenologischen Philosophie, Zweites Buch*, Dordrecht: Kluwer Academic Publishers, 1991, p. 17 (abbreviated title: *Ideen II*).

29 *Ideen II*, p. 22.

30 Ibid., pp. 22–23.

31 *Ideen I*, p. 387.

6 Is phenomenological time real time?

1 *La Phénoménologie de la Perception*, Paris: Gallimard, 1945, ii–iii. See also Levinas: "The realities obtained by physico-mathematical science borrow their sense from

operations which originate in the sensory." *Totalité et Infini*, The Hague: Martinus Nijhoff, 1980, p. 166.

2 Heidegger, *Prologemena zur Geschichte des Zeitbegriffs*, Frankfurt: Klostermann (Gesamtausgabe, vol. 20), 1994, p. 442; emphasis original.

3 Sartre, *L'Etre et le Néant*, Paris: Gallimard, 1943, p. 255.

4 Thus Levinas: "It is to Bergson that is due the merit of having freed philosophy from the spellbinding model of scientific time." *Ethique et Infini*, Paris: Fayard, 1982, p. 18.

5 'Time and eternity', *The Monist*, 1978, p. 7.

6 Merleau-Ponty, *L'Œil et l'Esprit*, Paris: Gallimard, 1964, pp. 9–10.

7 Nagel, *The View from Nowhere*, Oxford: Oxford University Press, 1986, p. 18.

8 For an account of the historical and contemporary influence of this principle, see T. Sorell, *Scientism: Philosophy and the Infatuation with Science*, London: Routledge, 1991.

9 Nagel rejects this dependency when he says: "my opposition to psycho-physical reduction is . . . fundamentally different from that of the idealist or phenomenological tradition" (*The View*, pp. 18–19). I doubt Nagel's position is as different from idealism as he thinks. For what is the difference between his view from nowhere and the impossible 'point of view of the whole' of neo-Hegelian idealism?

10 As prescribed by Carl Hempel's canonical statement: "All psychological statements which are meaningful, that is to say, which are in principle verifiable, are translatable into propositions which do not involve psychological concepts, but only the concepts of physics. The propositions of psychology are consequently physicalistic propositions." 'Data, reality and the mind–body problem", in H. Feigl and W. Sellars (eds), *Readings in Philosophical Analysis*, New York: Appleton-Century-Crofts, 1949, p. 378. For 'psychology', read 'phenomenology'.

11 J. Barbour, *The End of Time*, London: Weidenfeld & Nicolson, 1999, p. 14.

12 Ibid.

13 Heidegger, *Sein und Zeit*, Tübingen: Max Niemeyer, 1986 (1927), p. 18; hereinafter 'SZ'.

14 SZ, p. 333.

15 F. Waismann, 'Analytic-synthetic', in R.M. Gale (ed.), *The Philosophy of Time,* London: Macmillan, 1968, p. 60.

16 For an expansion on this with particular reference to Poincaré's conventional measurement, see G.J. Whitrow, *The Natural Philosophy of Time*, London and Edinburgh: Nelson, 1961, pp. 45–46.

17 Harris, 'Time and change', *Mind*, 1957, vol. lxvi, p. 234.

18 Ibid.

19 Kant, *Critique of Pure Reason*, trans. N. Kemp-Smith, London: Macmillan, 1990, p. 180.

20 Ibid., p. 181.

21 Ibid.

22 Ibid., pp. 183–184.

23 Ibid., p. 185; emphasis added.

24 *The Confessions,* Oxford: Oxford University Press, 1992, p. 237; hereinafter 'C'.

25 *Essais sur les Données immédiates de la conscience*, Paris: Presses Universitaires de France, 1967 (1889), pp. 58–59; hereinafter 'DI'.

26 *Durée et Simultananéité*, Paris: Presses Universitaires de France, 1968 (1922), p. 52.

27 Ibid., p. 60.

28 G. Frege, *The Foundations of* Arithmetic, Oxford: Basil Blackwell, 1978, p. 45.

29 Ibid., p. 53.

30 'The philosophy of Bergson', *The Monist*, 1912, vol. xxii, p. 334.

31 J.N. Findlay, 'Time: a treatment of some puzzles', in. A.G.N. Flew (ed.), *Logic and Language* (First Series), Oxford: Blackwell, 1978, pp. 47–48.
32 As was pointed out by A.O. Lovejoy in his early and still challenging article "The problem of time in French philosophy", *The Philosophical Review*, 1912, vol. 21, p. 60.
33 Mainly in *Die Grundprobleme der Phänomenologie*, Frankfurt: Klostermann (Gesamtausgabe vol. 24), 1989; hereinafter 'GP'.
34 See E. Martineau, '"Conception vulgaire et conception Aristotélienne du temps": note sur "Grundprobleme der Phänomenologie de Heidegger (para. 19)"', *Archives de Philosophie*, 1980, vol. 43, pp. 99–120.
35 *The Problem of Knowledge*, Harmondsworth: Penguin, 1971 (1956), p. 152.
36 *The Language of Time*, New York: Humanities Press, 1968, p. 67.
37 Paris: Editions de Minuit, 1986; hereinafter 'HPE'.
38 This canine example may give purchase on Franck's lapidarian remark: "as is phenomenally attested by sexuality, ... [m]y flesh is, in and of itself, *between* itself and what is other than itself" (HPE, 96). As that 'flight' out of itself, my flesh 'is not'(presented) and is spatializing.
39 D.H. Mellor, *Real Time*, (Cambridge: Cambridge University Press, 1987 (1981), p. 92.
40 *L'Etre et le Néant,* p. 27.
41 Heidegger, *Was ist Metaphysik?*, Frankfurt: Klostermann, 1986, p. 23.

Bibliography

Aristotle, *Physics*, in *The Basic Works of Aristotle,* New York: Random House, 1941.

Augustine, St., *Confessions,* Oxford and New York: Oxford University Press, 1992.

Ayer, A.J., *The Problem of Knowledge,* Harmondsworth: Penguin, 1956.

Barbour, J., *The End of Time*, London: Weidenfeld & Nicolson, 1999.

Bergson, H., *Essais sur les Données Immédiate de la Conscience*, Paris: Presses Universitaires de France, 1967.

——, *Matière et Mémoire*, Paris: Presses Universitaires de France, 1968.

——, *Durée et Simultanéité*, Paris: Presses Universitaires de France, [1922] 1968.

——, *Les Deux Sources de la Morale et de la Religion*, Paris: Presses Universitaires de France, 1969.

——, *La Pensée et le Mouvant*, Paris: Presses Universitaires de France, 1969.

——, *L'Evolution Créatrice*, Paris: Presses Universitaires de France, 1969.

——, *L'Energie Spirituelle*, Paris: Presses Universitaires de France, 1972.

Bernet, R. (ed.), *Edmund Husserl: Texte zur Phänomenologie des inneren Zeitbewusstseins (1893–1917)*, Hamburg: Felix Meiner Verlag, 1985.

Boundas, C., 'Deleuze–Bergson: an ontology of the virtual', in P. Patton (ed.), *Deleuze*: A *Critical Reader*, Oxford: Blackwell, 1996, pp. 81–106.

Broad, C.D., *Scientific Thought,* reprint, London: Humanities Press, 1969.

——, *Ethics and the History of Philosophy*, London: Routledge & Kegan Paul, 1952.

Brough, J. 'The emergence of an absolute consciousness in Husserl's early writings on time-consciousness', in F. Elliston and P. McCormick (eds), *Husserl: Expositions and Appraisals*, Notre Dame: University of Notre Dame Press, 1977.

——, *Edmund Husserl: On the Phenomenology of the Consciousness of internal Time*, London: Kluwer Academic Publishers (trans. of Husserl 1966), 1991.

Camus, A., *Essais*, Paris: Pléiade, 1965.

Christensen, F. (1974) 'McTaggart's paradox and the nature of time', *The Philosophical Quarterly,* 1974, vol. 24, no. 97, pp. 289–299.

Copleston, F., *A History of Philosophy*, vol. ix: *Maine de Biran to Sartre*, London: Search Press, 1975.

Dainton, B. (2000) *Stream of Consciousness*, London: Routledge.

Deleuze, G., 'La conception de la Différence chez Bergson', *Les Etudes Bergsoniennes*, 1956, vol. 4, pp. 79–112.

——, *Le Bergsonisme*, Paris: Presses Universitaires de France, 1966.

——, *Différence et Répétition*, Paris: Presses Universitaires de France, 1968.

——, *Logique du Sens*, Paris: Les Editions de Minuit, 1968.

Derrida, J., *La Voix et le Phénomène*, Paris: Presses Universitaires de France, 1967.

——, *L'Ecriture et la Différence*, Paris: Editions du Seuil, 1967.
Deutsch, D., *The Fabric of Reality*, London: Penguin Books, 1997.
Duval, R., 'La durée et l'absence', *Revue des Sciences Philosophiques et Théologiques*, 1981, vol. 65, pp. 521–572.
Eddington, A.S., *Space, Time and Gravitation*, Cambridge: Cambridge University Press, 1953.
Findlay, J.N., 'Time: a treatment of some puzzles', in A.G.N. Flew (ed.), *Logic and Language* (First Series), Oxford: Blackwell, 1978, pp. 37–54.
——, 'Time and eternity', *The Monist*, 1978.
Flew, A.G.N. 'The sources of serialism', in S.C. Thakur, *Philosophy and Psychical Research*, London: Allen & Unwin, 1976, pp. 81–96.
Franck, D., *Heidegger et le Problème de l'Espace*, Paris: Les Editions de Minuit, 1986.
Frege, G., *The Foundations of Arithmetic*, Oxford: Basil Blackwell, 1978.
Gale, R.M., 'Some metaphysical statements about time', *The Philosophical Review*, 1966, vol. lii, pp. 320–330.
——, (ed.), *The Philosophy of Time*, London: Macmillan, 1968.
——, *The Language of Time*, New York: Humanities Press, 1968.
Gallagher, S. *The Inordinance of Time*, Evanston, IL: Northwestern University Press, 1998.
Granel, G., *Le Sens du Temps et de la Perception chez E. Husserl*, Paris: Gallimard, 1968.
Harris, E., 'Time and change', *Mind*, 1957, vol. lxvi, pp. 233–241.
Heidegger, M. *Sein und Zeit*, Tübingen: Max Niemeyer, [1927] 1986.
——, *Was ist Metaphysik?*, Frankfurt: Klostermann, 1986.
——, *Die Grundprobleme der Phänomenologie*, Frankfurt: Klostermann (Gesamtausgabe vol. 24), 1989.
——, *Prologemena zur Geschichte des Zeitbegriffs*, Frankfurt: Klostermann (Gesamtausgabe vol. 20), 1994.
Hempel, C., 'Data, reality and the mind-body problem', in H. Feigl and W. Sellars (eds), *Readings in Philosophical Analysis*, New York: Appleton-Century-Crofts, 1949.
Howells, C., 'Qui perd gagne: Sartre and Derrida', *The Journal of the British Society for Phenomenology*, 1982, vol. 13, pp. 26–34.
Husserl, E., *Ideen zu einer reinen Phänomenologie und phänomenologischen Philosophie*, Erstes Buch, The Hague: Martinus Nijhoff, 1950.
——, *Zur Phänomenologie des inneren Zeitbewusstseins (1893–1917)*, The Hague: Martinus Nijhoff, 1966.
——, *Ideen zu einer reinen Phänomenologie und phänomenologischen Philosophie*, Zweites Buch, The Hague: Martinus Nijhoff, 1991.
——, *Cartesianische Meditationen und Pariser Vorträge*, The Hague: Martinus Nijhoff, 1973.
Kant, E., *Critique of Pure Reason*, trans. N. Kemp-Smith, London: Macmillan, 1990.
Kearney, R., *Dialogues with Contemporary Continental Thinkers*, Manchester: Manchester University Press, 1984.
Kolakowski, L., *Bergson*, Indiana: St. Augustine's Press, 1985.
Lacey, A.R., *Bergson*, London: Routledge, 1993.
Levinas, E., *De l'Existence à l'Existant*, Paris: Fontaine, 1947.
——, 'La réalité et son ombre', *Les Temps Modernes*, 1948, vol. 38, pp. 771–789.
——, *Théorie de l'Intuition dans la Phénoménologie de Husserl*, Paris: Vrin, 1970.
——, 'Vérité de dévoilement et vérité de témoignage', *Archivio di Filosophia*, Rome, 1972, pp. 101–110.

——, *En Decouvrant l'Existence avec Husserl et Heidegger*, Paris: Vrin, 1974.
——, *Autrement qu'Etre ou au delà de l'Essence*, The Hague: Martinus Nijhoff, 1978.
——, *Totalité et Infini: Essai sur l'Extériorité*, The Hague: Martinus Nijhoff, 1980.
——, *Ethique et Infini*, Paris: Fayard, 1982.
——, *Le Temps et l'Autre*, Paris: Presses Universitaires de France, 1989.
——, *De Dieu qui Vient à l'Idée*, Paris: Vrin, 1992.
Lovejoy, A.O., 'The problem of time in French philosophy', *The Philosophical Review*, 1912, vol. 21, pp. 322–343.
Lloyd, G., 'Time and existence', *Philosophy*, 1978, vol. 53, pp. 215–227.
Martineau, E., 'Conception vulgaire et conception aristotélienne du temps: Note sur *Grundprobleme der Phänomenologie* de Heidegger (para. 19)', *Archives de Philosophie*, 1980, vol. 43, pp. 99–120.
McTaggart, J.E., 'The unreality of time', *Mind*, 1908, vol. 68, pp. 457–484.
——, *The Nature of Existence*, Cambridge: Cambridge University Press, 1927.
Mellor, D.H., *Real Time*, Cambridge: Cambridge University Press, (1981) 1987.
Merleau-Ponty, M., *La Phénoménologie de la Perception*, Paris: Gallimard, 1945.
——, *Eloge de la Philosophie*, Paris: Gallimard, 1960.
——, *Le Visible et l'Invisible*, Paris: Gallimard, 1964.
——, *L'Œil et l'Esprit*, Paris: Gallimard, 1964.
——, *L'Union de l'Âme et du Corps chez Malbranche, Biran et Bergson*, Paris: Vrin, 1968.
Moore, F.C.T., *Bergson: Thinking Backwards*, Cambridge: Cambridge University Press, 1996.
Mullarkey, J., *Bergson and Philosophy*, Edinburgh: Edinburgh University Press, 1999.
——, (ed.), *The New Bergson*, Manchester: Manchester University Press, 1999.
Nagel, T., *The View from Nowhere*, Oxford: Oxford University Press, 1986.
Pears, D.F., 'Time, truth and inference', *Proceedings of the Aristotelian Society*, 1950–1951, vol. lii, pp. 1–24.
Prior, A.N., *Papers on Time and Tense*, Oxford: Oxford University Press, 1968.
——, 'The notion of the present', *Studium Generale*, 1970, vol. 23, pp. 245–248.
Rankin, K., 'McTaggart's paradox: two parodies', *Philosophy*, 1981, vol. 56, pp. 333–346.
Russell, B. 'The philosophy of Bergson', *The Monist*, 1912, xxii, pp. 321–347.
——, *History of Western Philosophy*, London: Allen & Unwin, 1961.
——, *The Analysis of Mind*, London: Allen & Unwin, 1971.
Sartre, J.-P., *l'Etre et le Néant*, Paris: Gallimard, 1943.
——, *La Transcendance de l'Ego,* Paris: Vrin, 1972.
——, 'Une idée fondamentale de la phénoménologie de Husserl: l'intentionnalité', *La Nouvelle Revue Française*, 1939, pp. 129–132.
——, *L'Imagination*, Paris: Presses Universitaires de France, 1969.
Shorter, M., 'Imagination', *Mind*, 1952, vol. lxi, pp. 528–542.
Smart, J.J.C., 'The river of time', *Mind*, 1949, vol. 58, pp. 483–494.
Sneddon, K., *Time*, London: Croom Helm, 1987.
Sokolowski, R., *The Formation of Husserl's Concept of Constitution*, The Hague: Martinus Nijhoff, 1970.
Sorrel, T., *Scientism: Philosophy and the Infatuation with Science*, London: Routledge, 1991.
Strasser, S., *Jenseits von Sein und Zeit*, The Hague: Martinus Nijhoff, 1978.
Ströker, E., 'Zeit und Geschichte in Husserl's Phänomenologie: zur Frage ihres

Zusammenhanges', in E.W. Orth (ed.), *Zeit und Zeitlichkeit bei Husserl und Heidegger*, Freiburg/Munich: Verlag Karl Alber, 1983, pp. 111–137.

Suter, R., 'Augustine on time with some criticisms from Wittgenstein', *Revue Internationale de Philosophie*, 1961–1962, pp. 378–389.

Talberg, I., 'Tenses and "Now', *Philosophical Quarterly*, 1963, vol. 13, pp. 298–310.

Waismann, F., 'Analytic-synthetic', in R.M. Gale (ed.), *The Philosophy of Time*, London: Macmillan, 1968, pp. 55–63.

Whitrow, G.J., *The Natural Philosophy of Time*, London and Edinburgh: Nelson, 1961.

Williams, D.C., 'The myth of passage', in R.M. Gale (ed.), *The Philosophy of Time*, London: Macmillan, 1968, pp. 98–116.

Wittgenstein, L., *Philosophical Investigations*, Oxford: Basil Blackwell, 1981.

——, *The Blue and Brown Books*, Oxford: Basil Blackwell, 1975.

Wood, D., *The Deconstruction of Time*, Evanston, IL: Northwestern University Press, 2001.

Wyschogrod, E., *Emmanuel Levinas: The Problem of Ethical Metaphysics*, The Hague: Martinus Nijhoff, 1974.

Yont, M., 'Two reversibilities: Merleau-Ponty and Derrida', *Philosophy Today*, 1990, vol. 34, pp. 129–140.

Zaner, R.M., *The Problem of Embodiment*, The Hague: Martinus Nijhoff, 1971.

Zwart, P.J., 'The flow of time', *Synthese*, 1972, vol. 24, pp. 133–157.

Index

The arrangement is letter-by-letter. Italic page numbers indicate figures and illustrations not included in the text page range. References to notes are prefixed by *n*.

eBooks – at www.eBookstore.tandf.co.uk

A library at your fingertips!

eBooks are electronic versions of printed books. You can store them on your PC/laptop or browse them online.

They have advantages for anyone needing rapid access to a wide variety of published, copyright information.

eBooks can help your research by enabling you to bookmark chapters, annotate text and use instant searches to find specific words or phrases. Several eBook files would fit on even a small laptop or PDA.

NEW: Save money by eSubscribing: cheap, online access to any eBook for as long as you need it.

Annual subscription packages

We now offer special low-cost bulk subscriptions to packages of eBooks in certain subject areas. These are available to libraries or to individuals.

For more information please contact webmaster.ebooks@tandf.co.uk

We're continually developing the eBook concept, so keep up to date by visiting the website.

www.eBookstore.tandf.co.uk